ENGINEERING EDUCATION

Designing an Adaptive System

Board on Engineering Education
Commission on Engineering and Technical Systems
Office of Scientific and Engineering Personnel
National Research Council

NATIONAL ACADEMY PRESS
Washington, D.C. 1995

The project that is the subject of this report was approved by the governing board of the National Research Council, whose members are drawn from the councils of the National Academy of Sciences, the National Academy of Engineering, and the Institute of Medicine. The members of the committee responsible for the report were chosen for their special competences and with regard for appropriate balance.

This report has been reviewed by a group other than the authors according to procedures approved by a Report Review Committee consisting of members of the National Academy of Sciences, the National Academy of Engineering, and the Institute of Medicine.

This study by the Board on Engineering Education was conducted under National Academy of Sciences/National Research Council's Cooperative Agreement (No. OSR-9344774) with the National Science Foundation. Additional contributors were the National Aeronautics and Space Administration, U.S. Department of Energy, National Academy of Engineering, The Boeing Company, and Xerox Corporation.

Library of Congress Catalog Card Number 95-69924

International Standard Book Number 0-309-05278-5

Copies of the report are available in limited supply from:

Board on Engineering Education
2101 Constitution Avenue, NW
Washington, D.C. 20418
202-334-3505
beed@nas.edu

Copies are available for sale from:

National Academy Press
2101 Constitution Avenue, NW
Box 285
Washington, D.C. 20055
800-624-6242
202-334-3313 (in the Washington Metropolitan Area)

Printed in the United States of America

BOARD ON ENGINEERING EDUCATION

DOROTHY S. ZINBERG, Lecturer, Public Policy Center for Science and International Affairs, Kennedy School of Government, Harvard University, Cambridge, Massachusetts

Liaison Members to the Board on Engineering Education

Commission on Engineering and Technical Systems:

NAOMI F. COLLINS, Executive Vice President, NAFSA: Association of International Educators, Washington D.C. (from July 1, 1994)
JAMES J. SOLBERG, Professor of Industrial Engineering, Purdue University (until June 30, 1994)
PAUL E. TORGERSEN, President, Corporate Research Center, College of Engineering, Virginia Polytechnic Institute and State University (until June 30, 1994)
GEORGE L. TURIN, Vice President, Teknekron Corporation, Menlo Park, California (until June 30, 1994)
ROBERT V. WHITMAN, Professor Emeritus of Civil Engineering, Massachusetts Institute of Technology, Cambridge (from July 1, 1994)

Staff

MARY KAYE BENNETT, Staff Assistant (until February 1994)
DUNCAN BROWN, Consultant/Writer (until October 1993)
ALAN E. FECHTER, Executive Director, Office of Scientific and Engineering Personnel (until October 1994)
COURTLAND S. LEWIS, Consultant/Writer (from July 1993)
KERSTIN B. POLLACK, Board Acting Director

Foreword

Since the early decades of this century, when engineering programs became well established at many U.S. universities, engineering leaders in academe and industry have conducted periodic evaluations of the path that engineering education ought to take. The "Wickenden report" of 1930 (SPEE, 1930); the two "Hammond reports," *Aims and Scope of Engineering Curricula* (SPEE, 1940) and *Engineering Education After the War* (SPEE, 1944); the "Grinter report" of 1955 (ASEE, 1955), and the 1985 report of the National Research Council's Committee on the Education and Utilization of the Engineer (the "Haddad report"; NRC, 1985) were all landmark studies of the past that contributed a strong sense of "where we are now" and "where we ought to go" in engineering education. Because they were authoritative, their recommendations were often heeded when decision makers in universities and government considered policy choices affecting program directions, curricula, funding, and faculty advancement.

However, one might argue that, at least in some senses, none of these reports was truly revolutionary. To a great extent, they described and reinforced unchanging principles that are basic to engineering education. It is startling to read them and recognize the consistency of many of their themes across the decades:

- the need for strong grounding in the fundamentals of mathematics and the physical and engineering sciences;
- the importance of design and laboratory experimentation;
- a call for more attention to the development of communication and social skills in engineers;

- the need for integration of social and economic studies and liberal arts into the curriculum;
- the vital importance of good teaching and attention to curriculum development; and
- the need to prepare students for career-long learning.

The various reports differ mainly in the relative weight accorded these themes. However, they also reflect changes in the fundamental political, economic, and social circumstances governing each period. Thus, the Wickenden report reflected the rapid expansion of large, technology-based industrial organizations; the Hammond reports reflected both the explosion of technologies and the exigencies of a World War; the Grinter report was a reaction to the wartime demonstration that engineers required better grounding in mathematics and the physical sciences; and the Haddad study responded to sharply declining engineering enrollments and sharply increasing industrial competition from overseas.

The same holds true for the current effort of the National Research Council's Board on Engineering Education (BEEd), reported here. The central themes remain, but the emphases among them and the specific terms with which they are approached are different. What prompted this particular study? Partly it is that the environment for engineering is different—even from that of the mid-1980s, in some critical respects. Chief among the new factors are the end of the Cold War and reduction in the defense budget; a persistent worldwide economic challenge, with major restructuring of business and industry to meet global competition; the ubiquitous and rapidly evolving applications of information technologies; a strong growth of minority and immigrant populations in the United States without concomitant representation in engineering; the entry of large numbers of women into the workforce, also without concomitant representation in engineering; and a widening recognition of the responsibility of engineers to consider the social and environmental impacts of their work.

Government programs also drive change. The National Science Foundation (NSF), which traditionally has focused on support of research and graduate education, has a mandate to support undergraduate and precollege education in science, engineering, and mathematics. In the NSF's Directorate for Engineering, several engineering education coalitions are pursuing a fundamental restructuring of parts of the undergraduate engineering curriculum. The Engineering Research Centers consider education a vital part of their mission. Outreach programs sponsored through several NSF directorates, particularly the Directorate for Education and Human Resources, offer engineering educational opportunities to women and minorities, the disabled, students from small colleges and non-engineering colleges, and high school students and teachers. The Clinton Administration's National Science and Technology Council, through its Committee on Education and Training, has developed a five-year strategic plan for science, mathematics, engineering, and technology education under which the efforts of all federal agencies will be coordinated. (The NSF efforts are integral to the council's strategic plan, as are similar programs at the Department of Energy, the Department of Education, and the National Aeronautics and Space Administration.)

The opening paragraphs of this foreword suggested that many of the problems in engineering education are perennial problems—"the more things change, the more they stay the same." Given the changes just described, a basic question is whether this is actually true today; that is, will engineering practice remain more or less the same in the future, or will it require radical rethinking of educational content and process to reflect the nature of new knowledge and the changing modes of its transmission, the globalization of technology, the changing nature of engineering jobs and career patterns, and the changing nature of the university itself. The BEEd has come to the conclusion that, in many areas, major change in the engineering education system is indeed necessary if it is to meet the needs of the nation and the world in the coming century. I agree with this assessment, and I urge your attention to this report. Coming as I do from the industrial sector, I also wish to issue a special call to the nation's industrial leaders to recognize the responsibility they have to help reform and sustain engineering education.

Norman R. Augustine, *Chairman*
National Academy of Engineering

The National Academy of Sciences is a private, nonprofit, self-perpetuating society of distinguished scholars engaged in scientific and engineering research, dedicated to the furtherance of science and technology and to their use for the general welfare. Upon the authority of the charter granted to it by the Congress in 1863, the Academy has a mandate that requires it to advise the federal government on scientific and technical matters. Dr. Bruce M. Alberts is president of the National Academy of Sciences.

The National Academy of Engineering was established in 1964, under the charter of the National Academy of Sciences, as a parallel organization of outstanding engineers. It is autonomous in its administration and in the selection of its members, sharing with the National Academy of Sciences the responsibility for advising the federal government. The National Academy of Engineering also sponsors engineering programs aimed at meeting national needs, encourages education and research, and recognizes the superior achievements of engineers. Dr. Robert M. White is president of the National Academy of Engineering.

The Institute of Medicine was established in 1970 by the National Academy of Sciences to secure the services of eminent members of appropriate professions in the examination of policy matters pertaining to the health of the public. The Institute acts under the responsibility given to the National Academy of Sciences by its congressional charter to be an adviser to the federal government and, upon its own initiative, to identify issues of medical care, research, and education. Dr. Kenneth I. Shine is president of the Institute of Medicine.

The National Research Council was organized by the National Academy of Sciences in 1916 to associate the broad community of science and technology with the Academy's purposes of furthering knowledge and advising the federal government. Functioning in accordance with general policies determined by the Academy, the Council has become the principal operating agency of both the National Academy of Sciences and the National Academy of Engineering in providing services to the government, the public, and the scientific and engineering communities. The Council is administered jointly by both Academies and the Institute of Medicine. Dr. Bruce M. Alberts and Dr. Robert M. White are chairman and vice chairman, respectively, of the National Research Council.

Preface

The Board on Engineering Education (BEEd) is charged with identifying significant issues in engineering education; facilitating communication about engineering education needs among academic, industrial, and government leaders; developing long-term strategies for engineering education in the context of rapidly changing circumstances, technologies, and demands; formulating timely policy recommendations; and stimulating actions to implement the strategies and policy recommendations. To that end, in 1991 the BEEd embarked on an effort to:[1]

- identify the critical challenges facing U.S. engineering education today;
- present a vision of engineering education for the future;
- develop a plan for meeting the challenges; and
- stimulate a nationwide effort to implement the plan.

The board's goal in this effort is to achieve an engineering education system that reflects the needs and realities of the United States and the world of the twenty-first century.

As a first step in that direction, following a series of meetings at which the viewpoints of a wide range of organizations and individuals interested in engineering education were heard, the BEEd prepared a working paper (NRC, 1993) that provided a preliminary framework for discussing policy, programmatic, and budgetary alternatives. The

[1]See Appendix A for complete task statement.

BEEd presented the working paper in four regional symposia to engineering faculty, administrators, policy makers in industry and government, and representatives of both professional societies and student groups.[2] Through the symposia discussions on a regional and national basis, involving many of the nation's 311 engineering schools, its professional engineering societies, and state and federal agencies, the board hoped to develop a consensus document setting forth plans for addressing the pressing issues described in the working paper. Following the four symposia, the board analyzed all comments voiced during the symposia as input to its further deliberations, which culminated in this report.

Thus, virtually all sectors of the nation's engineering education community have participated in the development of this report and the actions it recommends, which are aimed at a realization of the BEEd's vision for an engineering education system appropriate to the next century. The board has not attempted to prioritize the many recommended actions; such an exercise would be not only difficult but also highly subjective. Instead, what was considered to be a more reasonable approach was taken by dividing the actions into two categories: those relevant to *all* institutions and "other possible actions for consideration." Also, four areas are singled out in Chapter 1 as high-priority actions.

I would point out that there are two key themes in this report that may distinguish it from other recent reports on engineering education. First, there is a broad recognition of the external context, national and increasingly worldwide, within which engineering education is conducted and of the fact that the culture of engineering education *must* adapt to that changing context. The second, related theme is the Beed's strong belief that engineering education institutions must evaluate themselves in the context of a shared vision of the future of the engineering education system, then determine which elements of that vision can be framed as objectives that are consistent with their particular institutional mission, and finally make the necessary changes to achieve those objectives. Thus, if there is a simple catch-phrase to describe our call to action, it is this: *"think globally, act locally!"*

On behalf of the BEEd, I would like to express my appreciation to the many individuals who contributed to this extensive study and who participated in the preparation of this report. Literally hundreds of

[2]The symposia and their participants are listed in Appendix B. Other contributors to the study, including presenters at smaller, topic-focused colloquia held by the BEEd, are listed in Appendix C.

people demonstrated their interest in the future of engineering education through active participation in meetings, symposia, and colloquia convened by the BEEd. (They are all listed in appendices B and C.) From the beginning of its deliberations, the BEEd has striven to ensure that its report would reflect as wide a spectrum as possible of the views of the engineering education community. Consequently, this is truly a consensus document; the ideas and beliefs of those many participants inform the report throughout.

A special thanks is extended to Charles M. Vest, who chaired the BEEd's Report Development Committee. A critical review process overseen by the National Research Council's Report Review Committee contributed to the refining of the report.

I would like in addition to acknowledge the valuable contributions of the National Research Council's Archie L. Wood, Executive Director of the Commission on Engineering and Technological Systems, and Alan E. Fechter, Executive Director of the Office of Scientific and Engineering Personnel. Finally, the BEEd gratefully acknowledges the excellent support provided by its staff members: Kerstin B. Pollack, our able Acting Director throughout the study; staff assistant Mary Kaye Bennett; consultant Duncan Brown, who drafted a white paper to assist the board in its deliberations; and consultant Courtland S. Lewis, whose work in synthesizing the material derived from the board's deliberations was indispensable. Without their assistance, this report would not have been completed successfully.

Karl S. Pister, *Chair*
Board on Engineering Education

Contents

ENGINEERING EDUCATION

Designing an Adaptive System

1

The Board's Message

Since 1991 the National Research Council's Board on Engineering Education (BEEd) has been taking stock of issues in engineering education, listening to the concerns of scores of educators and employers of engineers, and considering the future of this vital enterprise. In the course of that study, the board has identified many aspects of the education enterprise that must be improved.[1] However, as the millennium approaches, no single concept or action is evident that can bring to engineering education the fundamental changes the board believes are needed.

Expansion of the nation's population and a growing demand for technology in the mid-1800s yielded the idea of land grant institutions incorporating engineering experiment stations, as codified in the Morrill Act and the Hatch Act. The experience gained in weapons system development and precision manufacturing during World War II highlighted the need, reflected in the "Grinter report" of 1955 (ASEE, 1955), to provide a sound scientific base for the education of engineers. These were specific responses to a clearly defined need.

The end of the Cold War has produced a different situation. While it certainly has had an impact on the engineering profession—and will continue to do so—it is but one more major change added to the astounding development of information technology, the rigors of global economic competition, the challenges of environmental protection, problems with an

---

[1] The BEEd did not attempt to rank-order its many recommended actions; such an exercise would be not only difficult but also highly subjective. Instead, a more reasonable approach was taken by dividing the proposed actions into two categories: those relevant to *all* institutions and "other possible actions for consideration." Also, four areas are singled out in this chapter as high-priority actions.

aging infrastructure, the accelerating diversity of the nation's population, and other great technical and social transformations.

In all these influences, the common denominator is complexity and rapid change; this is the challenge faced today in engineering education. *The BEEd's deliberations have led the board to conclude that there is no simple, universal prescription for dealing with complexity and constant change.* Rather, there must be many responses, all individualized and tailored to local circumstances. Yet these localized responses must be made in the light of a global perspective, shared by all engineering educators and enlightened by input from employers and graduates, of the broader purposes, goals, and desired outcomes of engineering education. The nation's engineering institutions must together make up the core of a robust system, deliberately seeking to educate students so that they will attain the characteristics described in the board's "vision for the twenty-first century" (see Chapter 2).

To meet the challenges that the nation faces, *each engineering college or school*[2] *should enter a period of experimentation, monitored by self-assessment and feedback from industry,* that is characterized by a willingness to change and by open, active communication across the engineering community. This process will likely reveal many needed actions. *The BEEd believes that one of the highest-priority actions within many engineering schools is to align the faculty reward system more fully with the total mission and purpose of the institution.* The reward system at each institution must ensure a proper balance among teaching, research, service, and professional activities to support the institutional mission. Institutional economic pressure must not be permitted to take priority in establishing this balance.

The BEEd anticipates that another high-priority item emerging from experimentation and self-assessment by engineering schools will be a recognition of the need to reform the undergraduate engineering education curriculum. The undergraduate educational experience establishes the professional orientation and knowledge base for the vast majority of the nation's engineers. It *must* impart to students as many as possible of the characteristics described in the BEEd's vision. Several curriculum reform efforts are now under way in engineering schools and coalitions of schools across the nation. Institutions not already involved in such reform should monitor these activities and use them as models or catalysts for their own internal

[2] The committee recognizes that different terms are used in different institutions, but for the sake of brevity the term "engineering school" will be used throughout this report to refer to the largest organizational unit for engineers within the university.

reform efforts. (Reforms also are needed in graduate engineering education; but the primary focus of this report is on undergraduate education, since it builds the base for future strengthening of graduate education.)

Also important in this regard, the BEEd believes, is the need to seriously consider alternatives to the standard four-year bachelor's degree. Many now recognize that four years is no longer enough time for the formal education of an engineer about to enter a lifelong career of professional practice—even assuming a commitment to continuous education after entering practice.

Table 1-1 presents the key ideas contained in this report. The left-hand column summarizes desired characteristics of the system and its output (primarily engineering graduates), as described in the BEEd's vision for the twenty-first century (see Chapter 2). To achieve these goals will require the actions delineated in the "call to action" that appears in the final chapter (Chapter 5) of the report; these actions are summarized in the second column of the table. Finally, the table identifies the sectors that would necessarily be involved as agents in carrying out these actions. The actions and sectors are spelled out in detail in Chapter 5.

Actions for *all* institutions include the following:

- Conduct institutional self-assessment.
- Redress imbalances in the faculty incentive system.
- Improve teaching methods and practices.
- Ensure that the curriculum supports the institution's strategic plan.
- Expand beneficial interactions and outreach.

There are other possible actions for consideration, which are grouped in Chapter 5 according to the type of organization for which the action is recommended. Such organizations include institutions, industry, professional societies, government, government–industry–university cooperatives, accrediting authorities, and other groups of the engineering community.

TABLE 1-1 Achieving BEEd's Vision of the Engineering Education System—Actions and Agents

VISION: DESIRED CHARACTERISTICS OF THE EDUCATION *SYSTEM*	ACTIONS[a]	AGENTS[b]	Page No.
System is highly adaptable and flexible	• **Each institution must conduct a self-assessment and self-evaluation**	1	2, 3, 14, 18, 42-43, 44-46
	• **Conduct periodic evaluation and obtain feedback on performance of system and its outputs (engineers)**	1	2, 44-46
	• **Establish/improve coordination with rest of university**	1	33, 43, 50
	• Consider "modularizing" the curriculum	1, 4	23, 48
	• Work with Accreditation Board for Engineering and Technology toward more flexible accreditation criteria	1, 4	53, 54
	• Explore educational innovations and practices in other countries	1, 2, 3	54
Curriculum at each institution integrates fundamentals with early and broad exposure to engineering practice aspects, as well as with design	• **Pursue undergraduate curricular reform, including early exposure to "real" engineering and more extensive exposure to interdisciplinary, hands-on, industrial practice aspects, team work, systems thinking, and creative design**	1, 4, 5	2, 21-25, 48-49
	• **Monitor ongoing experiments in curricular reform and implement pertinent aspects, ensuring continued strong grounding in engineering science and math**	1, 4	2, 22-23, 49
	• **Employ on the faculty more engineers from industry and government with design and management experience**	1, 5	27, 50, 51
	• National Science Foundation should disseminate and implement results of the Engineering Education Coalitions as they become available	1 ,3	53

[a]Items in boldface are applicable to *all* institutions; items not in boldface should be considered for possible implementation by *some* institutions.

[b]LEGEND: 1 = Engineering school faculty and administration
2 = Professional societies
3 = Federal agencies
4 = Accreditation Board for Engineering and Technology and regional accrediting bodies
5 = Industry

TABLE 1-1 Continued

VISION: DESIRED CHARACTERISTICS OF THE EDUCATION *SYSTEM*	ACTIONS[a]	AGENTS[b]	Page No.
Offers a variety of paths to the B.S., M.S., and Ph.D. to provide for different combinations and types of knowledge and experience	• **Consider and implement, as appropriate, alternative paths to the undergraduate degree, including:** - **pre-professional "general engineering" degree** - **three- or four-year pre-engineering programs leading to a graduate engineering degree** - **cooperative degree** - **five-year bachelor's**	1, 4, 5	3, 16, 24, 48
	• **Consider and implement, as appropriate, alternative paths to the graduate degree, including:** - **practice-oriented M.S.** - **combined B.S./M.S.** - **industrial research and development track Ph.D. or D.Eng.** - **practice-oriented doctorate**	1, 4, 5	16, 48
	• Accreditation Board for Engineering and Technology should adopt, whenever possible, measurable performance- or output-oriented accreditation criteria	1, 4, 5	16
Offers a wide variety of opportunities and incentives for effective continuous education	• **Develop practice-oriented graduate study modules**	1, 2, 5	48
	• Remove barriers and provide incentives to engineers to pursue continuing education	5	15, 52
	• Adopt a sabbatical system to reward industrial employees with continuing education options	5	52
	• Societies and universities should collaborate in providing lifelong learning	1, 2	38
	• Societies should hold more education sessions at technical conferences	2	53
	• A federally supported coalition of university and industrial organizations should develop multimedia network(s) for continuing education	1, 2, 3, 5	53-54
Includes mechanisms to ensure diversity of students and faculty	• **Pursue diversity of the *student body* by: (1) improving access for all to engineering education; (2) conducting self-assessment of the diversity of the student body to identify needed corrective actions; (3) creating a positive, supportive climate that ensures racial, gender, and ethnic**	1, 3	16-17, 27-30, 49-50

TABLE 1-1 Continued

VISION: DESIRED CHARACTERISTICS OF THE EDUCATION *SYSTEM*	ACTIONS[a]	AGENTS[b]	Page No.
	diversity; (4) establishing formal commitments and incentives to balance faculty/student demographics; (5) improving articulation with community colleges and providers of continuing education; and (6) giving academic credit for specified "life experience"		
	• **Improve *faculty* diversity of race, gender, ethnic background, and age by: (1) altering the mix of faculty characteristics through self-initiated actions; and (2) employing more engineers from private industry and government with engineering design experience and management experience**	1, 3, 5	16-17, 30, 50
	• Develop a variety of faculty types and tracks, employing practitioners	1, 5	51
	• Fund fellowship programs and scholarships for women and minority engineering students	5	52
Educational experience is richer and is delivered with maximum productivity and cost-effectiveness	• **Provide incentives to encourage excellence in teaching, pedagogy, curriculum development, and multimedia teaching approaches**	1, 2, 3	3, 32, 47
	• **Develop and adopt criteria and practices for evaluating teaching effectiveness**	1, 2, 3, 4	31-32, 46-47
	• **Employ state-of-the-art teaching methods informed by cognitive science and reflecting changing learning styles, with expanded use of educational technology**	1	17, 25-27, 47
	• **Ensure greater participation by faculty in teaching undergraduates, emphasizing student-faculty interaction**	1	31, 47
	• **Create a positive, supportive climate for engineering students**	1	25-26, 47

[a]Items in boldface are applicable to *all* institutions; items not in boldface should be considered for possible implementation by *some* institutions.

[b]LEGEND: 1 = Engineering school faculty and administration
2 = Professional societies
3 = Federal agencies
4 = Accreditation Board for Engineering and Technology and regional accrediting bodies
5 = Industry

TABLE 1-1 Continued

VISION: DESIRED CHARACTERISTICS OF THE EDUCATION *SYSTEM*	ACTIONS[a]	AGENTS[b]	Page No.
	• **Employ on the faculty more practicing engineers with design and management experience who demonstrate good teaching abilities**	1	27, 50, 51
	• **Specialize the institution's program offerings to focus available resources**	1, 3, 4, 5	50-51
	• **Consider alternatives to tenure such as fixed-year contracts**	1, 2	51
	• **Document excellent teaching and teachers**	1	51
	• **Develop curricular models and instructional modules from inter-disciplinary building-blocks**	1, 3	23, 48, 51
	• **Release industry professionals to teach in universities for a limited period**	5	52
	• **Societies should honor faculty excellence in education**	2	53
	• **National Science Foundation could fund development of teaching tools for use by engineering educators**	1, 3	53
	• **Develop a nationwide instructional television network for undergraduate instruction**	1, 3, 5	54
Offers a wide diversity of educational approaches across different institutions	• **Consider and implement, as appropriate, alternative paths to undergraduate and graduate degrees**	1, 4	3, 16, 24, 48
	• **Consider graduate education reform as an integral part of graduate track B.S. and joint B.S./M.S. program reforms**	1, 4	48
	• **Pursue appropriate undergraduate curricular reform**	1, 4, 5	2, 48-49
	• **Develop "new collegiality"–a shared sense of mission and purpose for the faculty and the institution**	1	33, 50
	• Specialize the institution's program offerings to focus available resources	1, 3, 4, 5	50-51

TABLE 1-1 Continued

VISION: DESIRED CHARACTERISTICS OF THE EDUCATIONAL *OUTPUTS*	ACTIONS[a]	AGENTS[b]	Page No.
Engineers are versatile, able to identify and solve problems	• **Pursue undergraduate curricular reform, including early exposure to "real" engineering and providing for more extensive exposure to interdisciplinary, hands-on, industrial practice aspects, team work, systems thinking, and creative design**	1, 4, 5	2, 16, 48-49
	• **Establish mechanisms to provide faculty members with greater exposure to engineering practice**	1, 3, 5	47-48
	• National Science Foundation should disseminate and implement results of the Engineering Education Coalitions as they become available	1, 3	23
U.S. engineers compete well in rapidly changing global markets	• **Pursue undergraduate curricular reform, including early exposure to "real" engineering and providing for more extensive exposure to interdisciplinary, hands-on, industrial practice aspects, team work, systems thinking, and creative design**	1, 4, 5	2, 16, 48-49
	• **Experiment with ways to expose students to the internationalization of industrial competitiveness and technology development**	1,4	24-25, 49
	• **Establish mechanisms to provide faculty members with greater exposure to engineering practice**	1, 3, 5	47-48
	• Become more international in institutional orientation and programs	1	51
Engineers possess better communications skills, a penchant for collaboration, and the capability for business and civic leadership	• **Create a positive, supportive climate for engineering students by emphasizing success and personal encouragement**	1	25-26, 47
	• **Pursue undergraduate curricular reform, including greater required exposure to principles of design, team projects, business, and liberal arts**	1, 4, 5	2, 15, 22, 48-49

[a]Items in boldface are applicable to *all* institutions; items not in boldface should be considered for possible implementation by *some* institutions.

[b]LEGEND: 1 = Engineering school faculty and administration
2 = Professional societies
3 = Federal agencies
4 = Accreditation Board for Engineering and Technology and regional accrediting bodies
5 = Industry

TABLE 1-1 Continued

VISION: DESIRED CHARACTERISTICS OF THE EDUCATIONAL *OUTPUTS*	ACTIONS[a]	AGENTS[b]	Page No.
Having the habit of lifelong learning and a knowledge of how to learn, engineers are prepared to function productively over the course of a career and, if they wish, to pursue successful careers in other fields	• Instill in students a desire for continuous and lifelong learning to promote professional achievement and personal enrichment	1	15, 24, 37-39, 49
	• Societies and universities should collaborate in providing continuing education	1, 2	38, 53
	• A federally supported coalition of university and industrial organizations should develop multimedia network(s) for continuing education	1, 3, 5	53-54
	• Establish an on-line electronic library of documents used to build modular tutorials for use by engineers and students	1, 3, 5	38-39, 54
Engineers are aware of the complex interrelationships between engineering and society	• **Ensure early exposure to "real" engineering and a sense of the role of responsible engineers in society**	1, 4, 5	16, 48
	• **Require the study of science, technology, and society (or equivalent) for undergraduates**	1, 4	49
	• **Employ on the faculty more engineers from industry and government with engineering design experience and management experience**	1, 3, 5	50, 51
Engineers understand how to design and develop complex technological systems	• **Expand the definition of creative research activity to incorporate measures of industrial relevance in assessing faculty performance**	1	46
	• **Employ on the faculty more engineers from industry and government with engineering design experience and management experience**	1, 3, 5	50, 51
	• Provide released time for faculty professional development, emphasizing participation in large, cross-disciplinary industry/government research projects	1, 5	51
Engineers are comfortable with working on cross-disciplinary teams	• **Experiment with such teaching techniques as cooperative learning and peer teaching**	1, 3	26, 47
	• **Reform the undergraduate curriculum to provide for more extensive exposure to cross-disciplinary industrial practice aspects and team work**	1, 4, 5	16, 48-49

TABLE 1-1 Continued

VISION: DESIRED CHARACTERISTICS OF THE EDUCATIONAL *OUTPUTS*	ACTIONS[a]	AGENTS[b]	Page No.
Most graduates have significant industrial contacts and exposure to hands-on aspects of engineering	• Find creative ways to utilize more industry engineers in teaching undergraduates	1, 5	47, 51
	• Establish mechanisms to provide faculty members with greater exposure to engineering practice	1, 3, 5	47-48
	• Reform the undergraduate curriculum to provide for more extensive exposure to hands-on, industrial practice aspects, team work, and creative design	1, 5	16, 48-49
	• Encourage engineering staff to participate in engineering education development activities	5	52
	• Speak to student groups, describing successful careers in industry	2, 5	52
	• Fund faculty fellowships, internships, and adjunct professorships	1, 5	52
Graduates reflect the nation's full range of gender, racial, and ethnic diversity	• **Pursue diversity of the student body by: (1) improving access for all to engineering education; (2) conducting self-assessment of the diversity of the student body to identify needed corrective actions; (3) creating a positive, supportive climate that ensures racial, gender, and ethnic diversity; (4) establishing formal commitments and incentives to balance faculty and student demographics; (5) improving articulation with community colleges and providers of continuing education; and (6) giving academic credit for specified "life experience"**	1, 3	16-17, 27-30 49

[a]Items in boldface are applicable to *all* institutions; items not in boldface should be considered for possible implementation by *some* institutions.

[b]LEGEND:
1 = Engineering school faculty and administration
2 = Professional societies
3 = Federal agencies
4 = Accreditation Board for Engineering and Technology and regional accrediting bodies
5 = Industry

TABLE 1-1 Continued

VISION: DESIRED CHARACTERISTICS OF THE EDUCATIONAL *OUTPUTS*	ACTIONS[a]	AGENTS[b]	Page No.
Increased public understanding of the nature and role in society of technology in general and engineering in particular	• Formally recognize the pursuit of technological literacy among the general population as part of the school's mission	1	36-37
	• Require all non-engineering undergraduates in the institution to take 1-2 survey courses on engineering and technology	1, 3	17, 51
	• To the extent possible, involve parents in K–12 technology literacy programs	1	35, 51
K–12 students and teachers are technologically literate and have a better understanding of engineering as a profession	• Establish, through statewide consortia, centers where K–12 teachers could acquire in-service training on teaching tools/topics supporting technological literacy	1, 2, 3	35, 51
	• Conduct a pre-service "summer school" for college students majoring in science/math education	1, 3	51
	• Encourage engineering faculty to establish partnerships with K–12 teachers	1, 2	52, 53
	• Encourage faculty to establish mentoring relationships with middle- and high-school teachers and students	1, 2	36, 52
	• Establish mechanisms by which some engineering graduates would teach K–12	1, 2, 3	50
	• Provide engineering instructional materials to K–12 schools and encourage practitioners to form partnerships with K–12 teachers	2, 5	51
K–12 students demonstrate improved competency in science and mathematics	• Take responsibility for improving K–12 science, math, and pre-engineering education	1	17, 33-36, 51
	• Support efforts to reform K–12 science and mathematics at the national, state, and local levels	1, 2	34-35, 54

2

Engineering at the Millennium: A New Vision

THE CHANGING WORLD OF ENGINEERING

As the twenty-first century nears, humanity's world is undergoing epochal change. The current century is giving way to a global economy in which market dominance is fragmented, widely distributed, and often short-lived. Human affairs, from the international to the personal, seem uncertain and transitory. Even the end of the Cold War, otherwise an entirely positive event, has removed tensions and imperatives that lent a sense of structure to U.S. national priorities for more than four decades.

A long-running global recession has thawed, not into the traditional economic boom but instead into a tepid and uncertain recovery. While some industrial sectors appear healthy, nevertheless it is a recovery that may be threatened by its unevenness between and within nations and by the great dispersion of purchasing power and personal demand across the global population. Increasingly austere federal budgets and restricted industrial expansion in many sectors have become chronic.

This circumstance of general instability and rapid change is having a profound impact on the practice of engineering in the United States. Restructuring, downsizing, mergers and acquisitions, curtailment of research and development, outsourcing, research collaboration, automation, offshore manufacturing, and offshore engineering (particularly of software) are all attempts to survive in the new economic environment or to capitalize on new opportunities; they all affect the demand for engineers and the demands placed on these engineers. At the same time, the ability of the federal government to support engineering research and graduate education at colleges and universities is diminished. And the retooling of the defense industry toward a focus on civilian technologies, with attendant declines in the defense budget, has brought turbulence and funding cuts to large sectors of engineering activity in both industry and academe.

However, engineering's role is more important than ever. With humanity's growing numbers and demands placing ever-increasing pressure on the resources of a shrinking world, creative and thoughtful use of engineering and technology will remain essential for solving the problems of energy, food, transportation, housing, health care, communication, manufacturing, education, and environmental protection and for fulfilling all the other requirements of modern life (NAE, 1991).

An explosion of technology is occurring. It is not an explosion that affects the outward look of the landscape, as occurred in the period from 1850 to 1950 with the emergence of factories, large bridges and dams, automobiles and airplanes, highway systems, electric power systems, telephones, and televisions. Instead, it is a revolution in the way things are designed, made, and controlled—in what they are made of and how they work.

This technological revolution is more subtle than past ones but just as pervasive and important in its impact on human life. Many of the technologies of today and tomorrow are internal rather than external in their function and impact; often they operate on a microscopic and molecular scale—or even invisibly, in the electromagnetic spectrum. New materials, for example, are opening the door to superconductivity, microelectronic robots, embedded sensors, human organ replacement, and ever-smaller and more powerful computers. Biotechnology, to take another example, holds enormous promise for producing a variety of small revolutions in medicine, agriculture, and other fields. Computerization and information technology are driving an accelerating increase in the productive organization of human enterprise, from manufacturing and business to entertainment, telecommunications, transportation systems, and the "information highway."

> This century will go down in history as the century of technology. . . In these almost one hundred years we developed the ability to move people and things between any two points on the globe in hours and to keep those points in instantaneous communication. We sow, reap, cook, communicate, manufacture, travel, clothe, entertain, educate, research, manage, cure, and kill by highly technological means.
>
> Simon Ramo (Ramo, 1988)

The changes affecting engineering are not just economic and technological but also social and cultural. In the United States, a demographic shift is occurring on a scale equal to those of the early twentieth century, as immigration from Latin America and Asia together with the growing population of resident Hispanic and African Americans alter the traditional U.S. view of "minority" and "majority." Along with the entry of large numbers of women into the workforce over the past two decades, these demographic shifts mean that engineering—traditionally a bastion of white males—must reshape many of its cultural foundations if it is to remain strong and relevant to the society it serves.

There is a widening recognition of the responsibility of engineers to consider the social and environmental impact of their work. In sharp contrast to the attitudes and practices that prevailed at mid-century and before, engineers today are required to design sustainable systems that consider as crucial inputs the environmental impact of their manufacture and use, their accessibility to people of diverse ethnicity and physical abilities, their safety, and their recyclability.

The means of delivery of engineering work are also changing; engineering work is no longer delivered solely through tangible products. Engineering services ranging from designs to software systems to technology assessments are delivered electronically around the world. Engineering education is very much an engineering service, and it, too, requires effective delivery systems.

Other changes are having a major impact on education generally. Television, computers, and video games appear to have modified significantly the ways that young people learn and are willing to learn. A number of societal factors have contributed to a loss of academic discipline that yields, among other things, fewer youngsters with an orientation toward and strong skills in mathematics and science.

All these aspects of the changing context of engineering affect engineering education in various ways. The engineering education system is feeling the stress of changing external conditions but has undergone only limited and sporadic changes in response; like all established enterprises, it resists large-scale change. But the time for such change is now at hand.[1] There is an urgent need for new vision and for taking stock to see where changes must be made if the system is to continue meeting the needs of the nation now and in the coming century.

A VISION FOR THE TWENTY-FIRST CENTURY

Engineering will be challenged as never before to shape the nature and quality of life in the twenty-first century. Engineering education will be at the forefront of the effort to meet that challenge.[2] The BEEd envisions a U.S. engineering education system that is highly adaptable to the demands of the future, producing well-rounded profes-

[1]Other authoritative groups are also recognizing this need. For example, in October 1994, the results of a major study, *Engineering Education for a Changing World*, were announced by the American Society for Engineering Education (ASEE, 1994), and the report *Restructuring Engineering Education* was issued by NSF in March 1995 (NSF, 1995).

[2]The use of "will" describes an ideal future state; it should not be read as an imperative dictate.

sional engineers able to work together efficiently in teams to identify and solve complex problems in industry, academe, government, and society.

Along with engineering itself, engineering education in the twenty-first century will have found new priorities and a new social role suited to the post–Cold War world. U.S. engineers will compete well in regional as well as global markets characterized by rapid technological change and intense competition. More of them will assume central roles in the management of academe, industry, and government, and all will have greater intellectual breadth, better communication skills, a penchant for collaboration, and a habit of lifelong learning. The teaching of these characteristics will apply to the education of future engineering faculty as well as to that of practitioners.

Given the rapidity of technological change, it is essential that the education system prepare students to function productively as engineers (whether in industry, government, or academe) over the full course of a career. Content-based learning alone must not drive engineering education. The primary aim will be to instill a strong knowledge of *how to* learn while still producing competent engineers who are well-grounded in engineering science and mathematics and have an understanding of design in the social context. Ideally, the education engineers obtain at the undergraduate level will be broad enough to provide a strong basis not only for a career in engineering but also for careers in other professions. This will give them the flexibility to pursue interests and opportunities in other fields—such as medicine, law, and management—where they can bring their technological perspectives to bear in useful ways, as well as to respond to changing market conditions for engineers.

Educational reforms at the graduate level likewise will provide students with the flexibility to function as faculty members, industry researchers, or product development team members and leaders. Graduate-level engineers will be comfortable with systems-oriented work and will be able to move with relative ease between different specialized areas of engineering research.

To ensure that engineers can continue to develop their knowledge and capabilities over a lifetime of practice, the system will offer a wide variety of opportunities for readily accessible and effective continuous education. Industry will establish clear incentives for practicing engineers to continuously improve their knowledge and competence.

DEFINITIONS

Science — The study of natural systems (including physical, mathematical, biological, behavioral, and social/economic systems) in order to discover new knowledge and improve human understanding of those systems.
Engineering Science — The study of natural and/or human-made systems and processes with a view to the eventual use of the knowledge obtained in engineered systems, products, processes, and services.
Engineering — The profession in which knowledge of the mathematical and natural sciences gained by study, experience, and practice is applied with judgment to develop ways to utilize, economically, natural and man-made materials and the forces of nature for the benefit of humankind.

Engineering education will endeavor to make students more aware of the complex interrelationships between engineering and industrialized society (including the natural environment), encouraging and preparing them to assume stronger and more visible roles—even leadership roles—as responsible engineers in society and as productive citizens (see, for example, Florman, 1987).

As part of that understanding of complexity, engineering graduates will have an orientation toward (and understanding of) the design and development of complex technological systems. To that end, they will be experienced and comfortable with working on cross-disciplinary teams whose members' primary expertise might encompass several engineering disciplines and the sciences, as well as business, law, and marketing, and in which each member has a basic understanding of the others' disciplines.

Central to the education of most engineers will be significant industrial contact and a strong educational exposure to the practical, hands-on aspects of engineering in both large, established corporations and small new ventures. The undergraduate curriculum at each institution will integrate the fundamentals of natural sciences, engineering science, and mathematics with early and broad exposure to these engineering practice aspects, as well as with creative design. All engineering students, regardless of their choice of career, will experience this integrated education. Such an experience is especially important in the education of future undergraduate and graduate engineering faculty, for the knowledge and perspectives of professors are transmitted to each new generation of engineers.

All these expectations, taken together, place enormous pressure on the concept of the four-year bachelors degree. Few students can absorb all the necessary technical and nontechnical knowledge as well as the requisite practical experience in four years (see, for example, Augustine, 1994). Thus, schools will experiment with and offer a variety of alternative paths to the bachelors degree, including those requiring more than four years. They will also offer alternative routes to graduate degrees, including practice-oriented doctoral degrees as a complement to (not a replacement for) the current research-oriented doctoral degrees. The role of accreditation in such experimentation will be a central one. Performance- or output-oriented accreditation will be developed to encourage the diversity in educational formats that the BEEd believes is vital for the future of engineering education.

In light of the rapidly changing demographic makeup of the nation and in view of the valuable contributions women and underrepresented racial and ethnic minorities can make, the participation of such individuals in all aspects of engineering will become substantially

THE PENDULUM SWINGS...

At the core of the BEEd's vision is a set of imperatives that have been recognized by a growing number of engineering educators in recent years. To take but one example, the 1989 Massachusetts Institute of Technology report *Made in America* called for the creation of a new cadre of students and faculty characterized by (1) interest in, and knowledge of, real problems and their societal, economic, and political context; (2) an ability to function effectively as members of a team creating new products, processes, and systems; (3) an ability to operate effectively beyond the confines of a single discipline; and (4) the integration of a deep understanding of science and technology with practical knowledge, a hands-on orientation, and experimental skills and insight (Dertouzos et al., 1989, p. 157).

greater. To provide full access to all who could benefit from an engineering education, engineering schools will institute mechanisms that ensure that the diversity of their student body and faculty reflects the changing demographics of the national and regional population from which they draw their students.

A very important development among engineering students and the population in general will be the growth of an enthusiasm about engineering and an appreciation of the central role it plays in society. Such positive attitudes will be formed early. Accordingly, efforts by engineering schools will aim at ensuring that precollege teachers and college-level teachers of non-engineering students understand the nature and role of technology as well as the requirements for engineering careers. To the extent possible, K–12 students will be imbued with greater knowledge of engineering and improved competence in mathematics and science, resulting in larger numbers of better-qualified and better-informed entrants into engineering study. They will understand clearly the distinctions between engineering and science. Engineering faculty willingly accept the responsibility to teach courses that provide engineers with an appreciation of the traditions of engineering and non-majors with an understanding of why and how engineering is practiced. Engineering educators' responsibilities will thus extend to explaining the nature of engineering to all who would profess to be educated, and the responsibilities of other educators will extend to incorporating requirements for technological literacy in their curricula.

The educational experience will be richer as well as more productive. Engineering educators will employ modern, enlightened methods in nurturing, teaching, and developing the students. Their teaching methods will benefit from the findings of cognitive science and will reflect the changing culture and learning styles of young people, who increasingly are visual learners—computer literate and computer dependent. The educators will become expert in the use of educational technologies and information systems to enhance their teaching effectiveness. Ways will be found to make the delivery of engineering education more cost-effective. (Some of the same techniques used by industry in its efforts to cut costs—restructuring, consolidation, collaboration, and electronic networking, for example—will be applied not only to the business functions of the university but also to some of the purely academic functions, such as the development of curricula and the delivery of courses.)

The vision of engineering education presented here cannot be static. Like the engineering education system itself, this vision must evolve to meet changing and unforeseen needs. The education system, including curricula, must continually change to reflect the emerging directions of the engineering profession and the evolving needs of the "customer"—the engineering student and practitioner. To that end, the BEEd considers *adaptability* to be an essential attribute of engineering education in the twenty-first century. Diversity of approaches is a crucial element of this adaptability. Engineering schools must be permitted to pursue these and future needs in their own varied ways, reflecting the variety of their student populations and of the regional industries, public works, and other determinants that shape their missions.

3

Engineering Education Today

SOME IMPORTANT STRENGTHS

The success of U.S. engineering education has long been recognized worldwide. There are 311 engineering schools in the United States,[1] which are open to academically qualified students from any country, class, race, or ethnic group. Top students from around the world vie to attend U.S. colleges and universities to study engineering. U.S. engineering education is solidly based on in-depth study of the natural sciences, engineering science, and mathematics, an approach recommended by the influential Grinter report in the 1950s (ASEE, 1955). Thus it is an education that is highly analytical and theoretical in nature, although in recent years increased attention has been given to instilling in undergraduates a better appreciation of design and other aspects of industrial practice.

Graduate education is particularly strong in many U.S. engineering schools, in part because it is based on a research enterprise that is, generally speaking, second to none. This research orientation in turn enriches the undergraduate curriculum and influences its character through lectures and textbook development by faculty who are at the frontier of their field of knowledge and through the use of graduate students as teaching assistants. Many schools have programs that also provide undergraduates with direct research experience. This orientation toward research and discovery is a major attraction for foreign

[1]This was the number of institutions in 1994 that had programs in engineering that were accredited by the Accreditation Board for Engineering and Technology. (The schools had 1,494 accredited degree-granting programs that year.)

students, who often take the knowledge gained back to their home countries and industries, where it is put to practical use in the global marketplace.

Despite these strengths, there are many areas where engineering education must improve if it is to remain the best in the world and better serve the needs of the nation.

AREAS NEEDING IMPROVEMENT

To attain the vision described in the preceding chapter will require changes in engineering education. Already, however, in each of the areas discussed below some pioneering engineering educators and institutions are pursuing new directions. Their approaches need to be disseminated, modified, and implemented more widely, and new approaches need to be tried and tailored to the circumstances and the nature of each institution. Some additional alternatives will be suggested in Chapter 5.

A number of the industrial participants at the BEEd symposia expressed the view that radical change is needed. Paul Rubbert, Chief of Aerodynamics Research at Boeing Company, said:

> A sense of urgency is missing. We need to recognize that the undergraduate process is *broken*, and cannot be fixed mainly by tinkering. Rather, it must be reinvented or reengineered. . .

Robert Richie, Director of University Affairs at Hewlett-Packard, agrees that "a complete reform and new mission is needed. . ." to produce needed changes.

Daniel Okun, Professor Emeritus of Environmental Engineering at the University of North Carolina at Chapel Hill, painted a troubling picture in a letter sent to the BEEd (Okun, personal communication, March 22, 1984). He noted that engineering is the only profession for which a four-year program of study is all that is required for professional status. As he pointed out:

- Prospective engineering students must make a decision to commit to engineering in the 11th grade; yet many of the brightest young people prefer to keep their career options open longer than that.
- A four-year undergraduate curriculum cannot provide engineering students with the same preparation for leadership as those who have enjoyed six or more years of higher education in preparation for other professions.

"I have become increasingly aware that in the average engineering project, the first 10 percent of the decisions made effectively commit between 80 and 90 percent of all the resources that subsequently flow into that project. Unfortunately, most engineers are ill-equipped to participate in these important initial decisions because they are not purely technical decisions. Although they have important technical dimensions, they also involve economics, ethics, politics, appreciation of international affairs, and general management considerations. Our current engineering curricula tend to focus on preparing engineers to handle the other 90 percent, the nut-and-bolt decisions that follow after the first 10 percent have been made. We need more engineers who can tackle the entire range of decisions."

D. Allan Bromley,
Dean of Engineering, Yale University,
Personal communication to the BEEd,
January 17, 1995

- Recognizing these limitations, many engineering students opt for graduate study in law or business; those who enter graduate engineering programs become more specialized in science and research, rather than in engineering.
- Given all the technological advances that have been made in engineering since mid-century, how can the same length of time now as then be adequate to prepare a student for a career in professional engineering?

Okun concluded by saying, "Many 'band-aid' solutions to these problems have been proposed and some acted upon, without much impact. Unless engineering educators are challenged to consider and adopt significant changes, *I fear that engineers in the future will be technicians, in the service of a better educated and prepared leadership drawn from other professions.*"

Undergraduate Curriculum

The one area in which change is needed most is the undergraduate engineering curriculum.[2] It is now widely believed that for several decades too much emphasis was placed on engineering science (analysis) at the expense of design (creative synthesis) and other aspects of the practice of engineering. Notwithstanding that students need a solid foundation in basic mathematics and physical science to formulate and solve problems, they also need much more exposure to the practice aspects of engineering. (Appendix D presents a description, developed by the BEEd, of the purposes and principles of a progressive new undergraduate curriculum.)

> "Engineering education needs to be a process that emphasizes synthesis and the integration of knowledge, and a much closer link among education, research, and professional practice."
>
> Francis C. Lutz,
> Dean of Undergraduate Studies,
> Worcester Polytechnic Institute,
> Personal communication to the BEEd,
> March 9, 1994

Many engineering educators and practitioners are asking, Does today's engineering curriculum adequately engage students? Does it prepare them to adapt to the changing demands of the current and future engineering workplace and life in a complex technological society? These general questions often take specific form, such as:

- Do students gain a real sense of engineering early enough to hold their interest?

[2]Graduate engineering education also is in need of reform. However, the BEEd focused primarily on undergraduate education as the area having the greatest influence on the competitiveness of U.S. industries, recognizing also that reforms here will build the base for future reforms at the graduate level.

- What should be taught as "fundamentals"?
- Does engineering education integrate the fundamentals well enough with design and experimentation?
- Is it sufficiently practice-oriented to prepare students to apply their knowledge quickly? (And should this be required in an undergraduate program?)
- Is individual achievement emphasized too strongly over teamwork?
- Does the curriculum instill a sense of the social and business context and the rapidly changing, global nature of engineering today and in the future?
- Is the curriculum updated frequently to reflect current and emerging technology and tools?
- Is the undergraduate educational experience broad enough and liberal enough to prepare students for possible entry into non-engineering professions, including general management?
- Does the curriculum instill a knowledge of how to learn and a desire to learn in a wide range of areas, both technical and nontechnical, over the course of a lifetime?
- How can the curriculum, along with requirements for an engineering degree, be structured so as to prepare students simultaneously for engineering practice and graduate study?

The essential question is: What minimum combination of fundamentals; skills; and acquaintance with problem formulation and solution, the process of design, and the nature of professional practice is required to satisfy the description of an engineer presented in the BEEd's vision?

"We introduced a new approach in the fall of 1991 that requires each engineering freshman to take two introductory engineering courses in the first year. These courses, offered by the six departments in the College of Engineering, emphasize problem-solving, hands-on, and design skills. The philosophy is to expose students early to "real" engineering, concurrent with fundamentals."

Edmond Ko,
Professor of Chemical Engineering,
Carnegie Mellon University,
Personal communication to the BEEd March 24, 1994

The National Science Foundation (NSF) has established several programs designed to promote comprehensive reforms in undergraduate engineering education. In 1988 it announced 10 awards in undergraduate curriculum development in engineering. The grants supported various approaches to improving undergraduate engineering learning, including experiments in planning, implementing, and disseminating new curricula (NSF, 1988).

One such initiative was Drexel University's experimental Enhanced Educational Experience for Engineering Students (E^4), which sought a comprehensive restructuring of the freshman and sophomore engineering curriculum in terms of objectives, subject matter, and instructional methods. The E^4 curriculum developed out of this effort stresses the unified foundations of engineering rather than the compartmentalized collection of principles, divorced from engineering applications, that occupy the first two years of conventional undergraduate study. It also promotes the development of communication skills and encourages vigorous, continuous, lifelong learning by exposing students to self-directed educational experiences and distance learning technologies. The university adopted the program throughout the College of Engineering in 1993–94 (Drexel University, 1992). Initial results have been extremely favorable: for example, 62 percent of students entering E^4 in fall 1989 received engineering degrees by the end of the 1994 summer term, compared with 32 percent of non-E^4 engineering students at Drexel during the same period of time (Drexel University, 1994).

In 1990, with the establishment of Engineering Education Coalitions, NSF supplemented sponsorship of curriculum development experiments on individual campuses with multi-campus dissemination of new curricula. Competitive awards are given to consortia of universities to participate in this program and support comprehensive curriculum reform at the engineering baccalaureate level. As of November 1994, a total of 58 colleges and universities were participating in eight coalitions, representing every region of the United States and every type of engineering school. NSF's goals in this program are to improve teaching, restructure the engineering curriculum, and increase the number of engineering bachelor's degrees awarded to women, members of underrepresented minorities, and people with disabilities. The program seeks to make engineering education more relevant and responsive to students by promoting creativity and the ability to learn independently (NSF, 1993).

A third NSF program, which began in 1991, was designed to encourage established engineering researchers in emerging fields to become involved in curriculum development. The Combined Research/Curriculum Development Program awards, as they are known, were each $400,000 over a three-year period, to be split evenly between research and curriculum development.

One goal of these government-funded curriculum development programs is to produce portable curriculum modules that can be shared among engineering schools nationwide—on-line or via videotape, text, television, and software—thereby increasing the dissemination of high-quality educational materials and reducing the workload on faculty. Many individuals believe that on-line tutorials in the form of "learning modules" hold much promise for the future of engineering education (McClintock, 1994).

Industry's efforts to reform undergraduate engineering education have been carried out generally on a smaller scale, with some

"For the student who needs a 'hands-on' experience and aims at a terminal B.S. degree, an appropriate model might be the German *Fachhochschule*."

"I have seen a well-run co-op program create lots of motivation and broaden the views of the students."

C.A. Desoer,
Professor Emeritus,
University of California, Berkeley,
Personal communication to the BEEd,
February 9, 1994

exceptions. For example, the American Electronics Association formed a Design to Deliver program, funded by several large corporations. In this three-year program, 15 companies are working with three universities to improve the product-quality and manufacturing emphasis of curricula and to help faculty members develop the knowledge and skills to carry out these improvements. Another significant effort toward reforming undergraduate engineering education was launched by The Boeing Company in 1994 (McMasters and White, 1994).

Discussion of the many elements of curriculum reform leads inevitably to a discussion of alternative paths to the bachelor's degree. It is not realistic to expect a single curriculum to prepare students for (1) engineering practice immediately after graduation, (2) graduate engineering study and research, and (3) graduate study in other fields. Instead, there is a need for a variety of options. For example, there could be three tracks to the bachelor's degree: a standard disciplinary degree, a "general engineering" degree offering the flexibility for pursuit of a master's degree in engineering or another professional field,[3] and a research-oriented track that is essentially the first four years of a research doctoral program. Various co-op (work-study) versions of the first two options might entail a heavier emphasis on industrial experience while making a longer program more affordable and improving the student's motivation and employment prospects. Each of the tracks should offer students the flexibility, in terms of knowledge or academic credits, to move to other tracks, and each should instill a knowledge of how to learn autonomously through exposure to distance learning and other media for obtaining continuous education.

The BEEd emphasizes that a sound engineering education is just the beginning of a lifelong educational experience. Perhaps the most important thing that a student can learn during the initial engineering education experience is how to continue learning on his or her own initiative. The distinction between education and training is a crucial one; knowing how to learn autonomously is a hallmark of *education*.

Finally, an aspect of U.S. engineering education that is often cited as desirable, but which is seldom addressed in the curriculum, is the

[3]Frank Schowengerdt, Vice President for Academic Affairs at the Colorado School of Mines, reports that the four-year general engineering degree is now the most popular option at the school, with 850 (in 1994) students majoring in an interdisciplinary degree accredited by the Accreditation Board for Engineering and Technology.

> "The focus should be on employing cooperative learning strategies and establishing classroom climates that encourage, not alienate or bore, the students. This does not mean lowered standards—quite to the contrary. I have completely changed my philosophy of "weeding out" students. . . . Now my students are learning much more, they are enjoying learning and are proud of their achievements (including learning communication skills); and hardly anyone drops out or fails, because I have set the target of "zero defects" and then provided the means for all students to succeed."
>
> Edward Lumsdaine,
> Dean of Engineering,
> Michigan Technological University,
> Personal communication to the BEEd,
> March 21,1994

need for graduates to have a sense of the global marketplace and the globalization of engineering. One factor of this need is that strong foreign competition in high-technology industries is still a relatively new phenomenon, and most faculty members have little direct experience with it. Another factor is that ways of addressing the issue—for example, learning foreign languages and providing for long- or short-term exchange of students—tend to be time-consuming and expensive. Other mechanisms, such as seminars presented by foreign-born faculty members (particularly those with industrial experience) and adjunct faculty from industry on aspects of this issue, might have value.

Teaching Styles and Methods

A widespread tradition in engineering education has been the "boot camp" approach, in which professors typically have made little effort to help students overcome the formidable demands placed upon them. The philosophy is that "if you are tough on them, the ones who survive have what it takes to be engineers." Thus, engineering education has traditionally been seen as a winnowing-out process. The old warning to entering students, "Look to your right and left; only one of you will graduate" is still valid. Only the most committed and competitive students survive for four years; overall retention rates for engineering programs are on the order of 65 percent (AAES, 1993, 1994).[4] Rigor and discipline are certainly necessary in engineering, but they are counterproductive when taken to such an extreme that many talented and capable students become alienated or simply lose interest (Seymour and Hewitt, 1994).

Static teaching methods do not help. The current environment for engineering education tends not to foster either good teaching or effective learning. It is generally recognized that today's young people, in contrast to their counterparts of a generation ago, are more oriented toward fast-paced, dynamic visual imagery. Yet engineering education often is still delivered as it was 50 years ago, by a professor standing in front of the lecture hall with a piece of chalk and a

[4]This is almost certainly a high estimate. It is based on a comparison of entering freshmen and graduates four years later and does not take into account freshmen with undeclared majors, students entering at later points from uncounted institutions such as two-year colleges, and other factors. More reliable data on retention do not exist.

pointer—or, more recently, an overhead projector—and relying on words and static symbols or drawings.

Teaching style can do much to communicate and reveal the excitement and allure of engineering, and even the lecturer can be quite effective if he or she is a talented presenter. But the lecture-hall format provides little or no opportunity for student–teacher interaction—especially for the mentoring, counseling, and nurturing that many students need. Most engineering faculty know little about how students learn; research on the cognitive processes of learning is relatively new. Very few engineering faculty possess any knowledge of this field. Yet it may hold promise for improving teaching and learning.

For example, many believe that highly participatory "active learning" methods are more effective for stimulating student interest and learning. One approach now coming into greater use is "cooperative learning," an instructional method that involves students working in teams to accomplish a common goal, under conditions that involve both positive interdependence (all members must cooperate to complete the task) and group accountability (each member is accountable for the entire final outcome). Inquiry laboratories, seminars taught by teams of teachers, and project-centered classes are other active learning strategies. Most emphasize teamwork—which emulates the way engineering is actually practiced—as opposed to the education of individual performers, which has been the traditional approach of engineering education.

The importance of teamwork as a vital component of engineering, whether in the classroom or in practice, can be dramatically enhanced by faculty teamwork in the delivery of education. The single-instructor classroom has its place, but team-teaching and shared responsibilities for course and curriculum development set an important example. Such team-oriented methods tend—through competition, cooperation, synergy, and peer pressure—to produce better teaching.

Nothing has been found that can replace strong, supportive, one-on-one interaction between a student and a faculty member. But many new educational technologies offer the possibility of making the delivery of engineering education more effective, more efficient, and more interesting. The potential for use of such

---

THE "CLASSICAL" ENGINEERING EDUCATION: METHODOLOGICAL PROS AND CONS

In terms of methodology and technology, the classical engineering education consists of a *teacher, blackboard, textbook, homework, and laboratory.*

The *advantages* of classical education are

- compulsion;
- credit;
- some adaptivity and customization;
- moderate attention factor;
- some interactivity;
- shared experience—friendship and misery;
- side channels and personal elements—jokes, etc.—;and
- continuity.

The *disadvantages* of classical education are

- it is paced to least common denominator,
- variability of teachers,
- it is often boring or poorly prepared,
- it is only moderately adaptive,
- modest use of graphics and visual material,
- teachers are often unprepared or unavailable for new subjects,
- laboratories are often obsolete and too expensive,
- blackboard handwriting is slow, and
- textbooks are often insufficiently explanatory.

---

technologies is growing rapidly but is still largely untapped. (The average engineer in industry utilizes a higher level of supporting technology than most academics do.) Several factors have combined in recent years to improve the potential of educational technologies. First is the increased availability and lowered costs of the technologies themselves, from videotape to personal computers to television satellite broadcasting. Data compression techniques (facilitating transmission of video images), the growing national information infrastructure, high-speed networks, multimedia conferencing, wireless digital communication, and hand-held computer notepads herald an even more exciting range of opportunities. Second, larger class sizes and a concomitant increase in demand for specialized courses suggest the potential usefulness of these technologies. Third, accompanying the growing demand is a scarcity of faculty to teach undergraduate courses, given budget constraints and the increasing pressure on faculty to focus on securing research grants and conducting cutting-edge research. Fourth, it can be anticipated that the advent of the "information highway" will alter students' styles of learning in the direction of these technologies.

Because the excellence and accessibility of U.S. graduate engineering education are recognized around the world, foreign nationals are very heavily represented in U.S. engineering schools. Their contributions as teaching assistants and faculty are vital, but some have trouble communicating in English, and others have been accused of bringing to the classroom inappropriate cultural attitudes—for example, regarding the roles of women and minorities (NRC, 1988).

Finally, it should be noted that one of the impediments to effective teaching of engineering is that so many engineering faculty lack sufficient contact with engineering practice. In the absence of such interaction, they are at a disadvantage in conveying to their students the excitement and opportunity that exists in professional engineering practice.

Diversity of Students and Faculty

Demographic change and the related issue of ethnic diversity pose major challenges to engineering education. The proportion of white college-age males in the national population, the group from which engineering has traditionally drawn its recruits, is declining steadily. Half of those retiring from the workforce by 2000 will be white males, but over 70 percent of new entrants to the workforce will be women, minorities, and immigrants. During the 1980s while the U.S. minority population grew by 35 percent, the white, non-Hispanic population grew only 2 percent (Vetter, 1992). At the same time, the number of

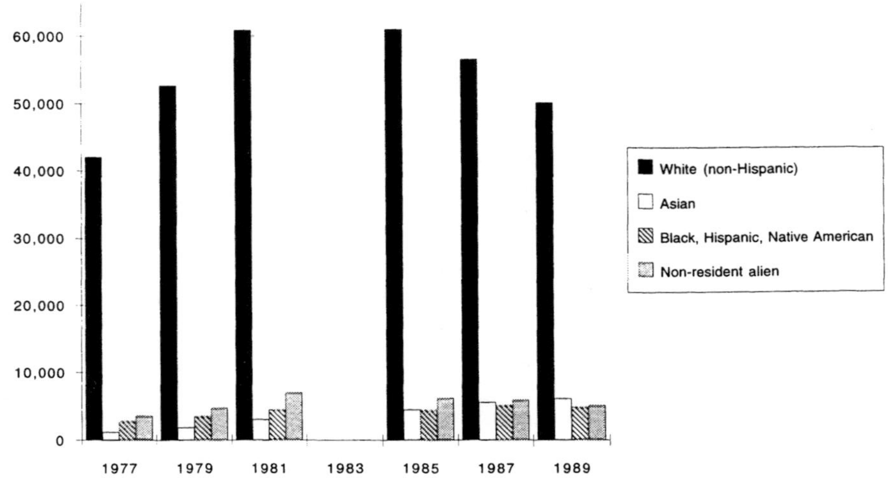

FIGURE 3-1 Engineering B.S. degrees, by race or ethnicity and residency status, selected years, 1977–1990 (National Science Foundation, 1992, p. 64).

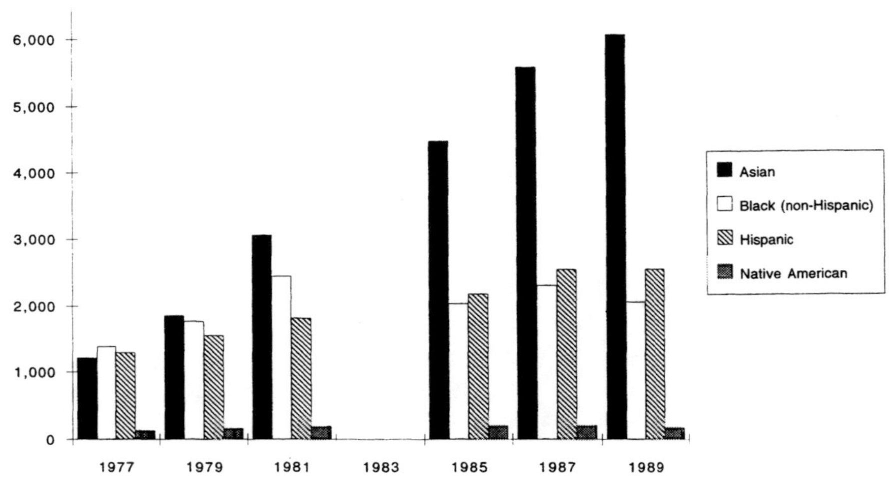

FIGURE 3-2 Engineering B.S. degrees to members of racial and ethnic minorities, selected years, 1977–1990 (National Science Foundation, 1992, p. 64).

white males achieving engineering degrees has declined sharply (Figure 3-1).

The number of racial and ethnic minority students receiving degrees in engineering increased somewhat during the 1980s (Figure 3-2), while the number of women declined from its peak in 1985 (Figure 3-3). Nevertheless except for male Asian Americans, who have made dramatic gains, none of these groups has approached full representation among engineering graduates. Today, women receive about 15 percent of B.S. engineering degrees, African Americans, Hispanics, and Native Americans—who together make up 27.5 percent of the college-age population—receive fewer than 8 percent of such degrees (NSF, 1992). Retention (the completion of a full academic program) is a special problem for minority students in engineering education; they represent more than 15 percent of first-year engineering students, but, as Figure 3-4 shows, more than half

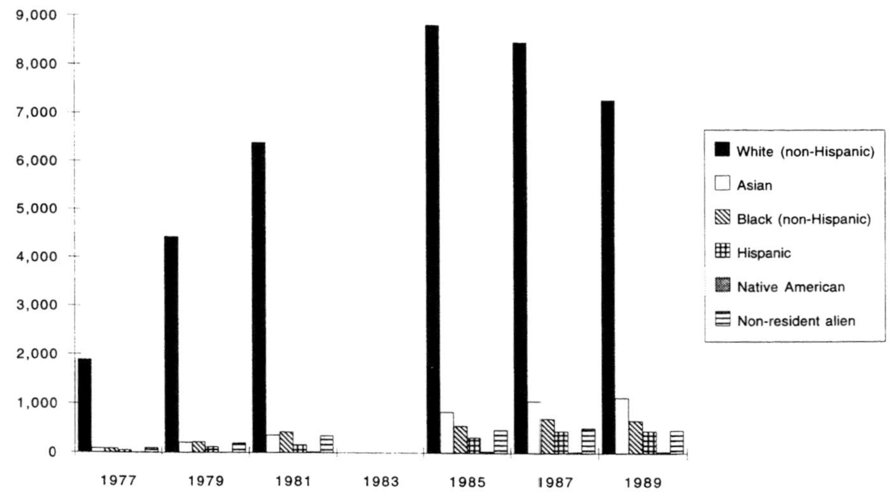

FIGURE 3-3 Engineering B.S. degrees to women, by race or ethnicity and residency status, selected years, 1977–1990 (National Science Foundation, 1992, p. 64).

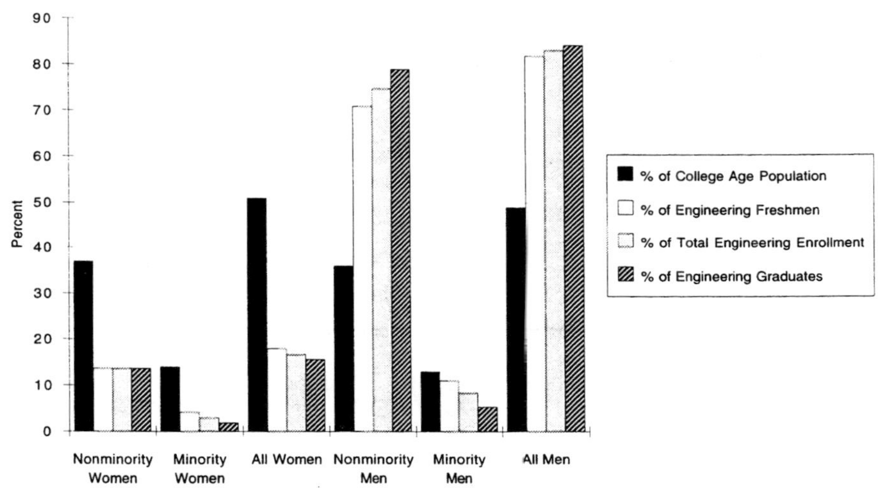

FIGURE 3-4 Representation of minority and nonminority groups in undergraduate engineering education and their representation in college age population, 1990–1991 (Campbell, 1992b).

drop out or switch to another major. For example (see Figure 3-4), minority men make up about 12 percent of entering students but only about 7 percent of graduates. Recent indications are that retention is only about 35 percent for African Americans and Native Americans and 45 percent for Hispanics, compared with roughly 65 percent for all freshmen and nearly 100 percent for Asians (bearing in mind that retention figures probably err on the high side). Anecdotal evidence suggests that the leading research universities are experiencing retention rates for minorities that are even lower than average.

The negative factors in engineering education described in previous sections appear to be magnified for women and minority students,[5]

[5]The BEEd recognizes that the experiences of (white) women and those of the various minority groups in engineering education differ considerably. These differences need to be taken carefully into account when designing ameliorative actions and programs.

who are often acutely aware of their underrepresentation and who may be even more put off than others by the boot camp atmosphere prevalent in undergraduate engineering education (Carmichael and Sevenair, 1991). Persistent anecdotal evidence points to discrimination—mostly unintentional or cultural but occasionally intentional—against underrepresented groups. According to Seymour and Hewitt (1994), the high number of foreign students and teaching assistants is part of the problem, as in some cases their cultural values impede positive interaction with women and minorities.[6]

Apart from retention, another very important factor is K–12 preparation. Female and minority students may be receiving the message, all through their early schooling, that a career in science or mathematics (or engineering) is not for them. Some aspects of the problem affect all students, regardless of race or gender. This issue is discussed in more detail in the section on K–12 preparation later in this chapter.

Most engineering faculties today remain bastions of white males, despite the changing demographics of their students and the even more rapidly changing demographics of the U.S. population as a whole. Although there has been an influx of non-white scholars from Asia and the Middle East, engineering faculties remain largely male. Many in the engineering community call for the engineering faculty of the future to be more diverse than that of today. "Diversity" has several different facets:

- diversity based on race, gender, and ethnic background;
- diversity of background in engineering practice, including design and management in industry and government; and
- diversity of academic background and orientation toward teaching, research, and professional practice.

Faculty characteristics do vary among institutions, reflecting in part differences in educational objectives. Nevertheless, greater faculty diversity—complemented by excellence—must be a goal for all institutions, not only to encourage equal access for all students but also to expose students to a wider spectrum of views as to what engineering is and how it is practiced, as well as to familiarize them

[6]The BEEd, in its regional symposia, addressed the question of the large population of foreign-born students and faculty and their effect on the engineering education system and presented a range of options for action. Many participants agreed that the appropriate course is to make no changes in the current system but rather to continue to seek the best students, regardless of their national origin. Therefore, this report does not raise the topic as an important issue.

with the composition of the society that is served by the practice of engineering.

Faculty Reward System

In engineering, and indeed across all academic disciplines, there is concern that the reward systems by which faculty performance is evaluated produce incentives that often lead faculty members onto a narrowly focused career path in academe. These incentives typically create a bias favoring research over undergraduate teaching while also discouraging mobility of faculty between academe, industry, and government. In effect, they may place a penalty on activities such as curriculum development, interactions with industry, outreach to precollege students, student advising, professional development, and other professorial functions designed to foster a more integrated academic community and a more well-rounded educational experience.

> "There is no fundamental dichotomy between research and teaching. Indeed, many would hold that good teaching over a career which spans 3-4 generations of new technology is impossible for one not engaged in research."
>
> John J. McCoy,
> Dean of Engineering,
> The Catholic University of America,
> Personal communication to BEEd,
> March 28, 1994

Nationwide, perhaps the most controversial aspect of the faculty reward system is the overemphasis on research at the expense of undergraduate teaching, which is seen at most schools to varying degrees.[7] While teaching usually has a prominent place in formal statements of faculty review criteria, it is often weighted lightly in faculty review processes. "Buying out" of teaching obligations with research dollars (being excused from teaching to conduct funded research) is an increasingly common practice in many institutions, encouraged by institutional financial pressure. This practice is detrimental to the quality of engineering education when carried too far and should be carefully monitored.

The roots of this situation lie in faculty attitudes toward teaching and in pressure from peers, academic administrators, and research funding agencies. Because many institutions today are operating with budgets that are far out of balance, faculty are expected to help make up the shortfall by securing research funds, thus reinforcing the emphasis on research. Another force tipping the balance toward research is that academic institutions, in making tenure and promotion decisions, generally find research quality a more straightforward

[7] Professor Robert Whitman, of the Massachusetts Institute of Technology, pointed out in a letter to the BEEd that the issue is not so much "research versus teaching," (since research is part-and-parcel of graduate education) as it is "research versus engineering" (since most students do not have sufficient opportunities to work with engineers who have experience in practice).

criterion to measure. Academe has developed accepted methods for evaluating the quality of research but has not developed comparable methods for evaluating teaching and professional service.

Recognition of this situation and its implications is growing. Many institutions are attempting to devise and institutionalize ways to recognize and reward effective teaching. (Massachusetts Institute of Technology's high-visibility program of internal MacVicker Faculty Fellowships is one example; another is Stanford University's Humanities and Sciences Dean's Award for Excellence in Teaching, which includes a base salary augmentation in addition to a cash award.) Some schools have instituted a non-tenure track faculty option that does not require teachers to pursue scholarly research, but this approach is highly controversial.

Changing the incentives will seriously challenge engineering faculties and academic administrators. At many institutions, a generation or more of faculty members have been hired and promoted primarily on the basis of their strengths in research. Efforts to change the incentives favoring research will be forced to face the fact that many faculty members consider research to be inherently more fulfilling and valuable than undergraduate teaching. In addition, the continued presence of faculty unions (which even extend to postdoctoral fellows and teaching assistants) may hamper efforts to change the incentive system. Finally, it will be necessary to develop a wider range of effective teaching assessment and evaluation methods and mechanisms.

The real issue, once these imbalances are rectified, is not whether research is favored over teaching but how to tie research to teaching in the most productive way or redefine research to include teaching (Boyer, 1991) and how to provide students with a broader vision of engineering than the collective scope of their professors' particular research areas can convey. Research and teaching are not antagonistic, and active involvement of undergraduates in frontier research is an excellent way to broaden their vision.

Flexibility and Adaptability

Engineering education tends to be conservative in both its pedagogical methods (including curriculum) and its institutionalized attitudes.[8] This conservatism produces a degree of stability (perhaps inflexibility is a more apt term) that results in a relatively slow response to external stimuli. A case in point might be an overempha-

[8]Perhaps the historical root of this conservatism is the responsibility for ensuring that engineering designs function safely and reliably.

sis on the production of engineering researchers, who compete for increasingly limited resources, at the expense of engineers advancing the state of engineering practice—especially in manufacturing and construction, where the need is great (White, 1991).

Given the many types of changes described earlier that are impinging on engineering, the engineering education system needs to become much more flexible and adaptable. Establishing interdisciplinary collaborations with science and liberal arts departments and business schools, in pursuit of both research and pedagogical developments, is an approach that could be useful (see, for example, Kapoor, 1994). It is possible that engineering schools will acquire greater flexibility through more extensive interaction with other educational units. Collaboration with industry and government also "ensures the vitality and relevance of engineering programs" and helps engineering students reach out more to the society around them (ASEE, 1994).

A New Collegiality

Collegiality, or the shared sense of mission, purpose, and values among the faculty, was a more common feature of academic institutions in the past. In the post–World War II era in engineering schools, this collegiality has tended to be eroded by trends such as larger institutional size; competitive grantsmanship; a loss of clarity about the role of engineering; and a narrower focus on the individual's social, political, and research interests (see Kerr, 1994, for example). A new collegiality in engineering departments and schools—which the BEEd believes is a vital element of responsible "institutional citizenship"—is essential if the actions and objectives of engineering education (e.g., the evaluation of teaching quality and curriculum renewal) are to be achieved. The new collegiality will be enhanced through organizing introductory courses, through professors lecturing in each other's courses—not only within departments and the engineering school but across the entire university—, and through including material in one's course that is outside one's field (necessitating collegial help), along with team teaching and peer evaluation of teaching.

> COLLEGIALITY AND TEACHING
>
> In a study of conditions within departments at 20 colleges and universities, Massy et al. (1994) found a high degree of collegiality being practiced in those "exemplary departments" that actively support undergraduate education. The distinctive characteristics of this collegiality include an emphasis on teaching, frequent interaction, tolerance of differences, generational and workload equity, peer evaluation, and consensus decision making. Collegial organizations, the authors stated, emphasize consensus, shared power, consultation, and collective responsibilities; they are communities in which status differences are de-emphasized and individuals interact as equals.

K-12 Preparation

The process of creating a successful engineering student begins early, in elementary school or even preschool. But the supply "pipeline," reaching from kindergarten through the senior year of high school (K–12), is not producing a sufficient flow of students who are informed about engineering and who are well-prepared and motivated to study engineering. It is not drawing from across the full breadth of the pool of potential engineers, and many young students do not obtain the knowledge and capabilities they need. In many cases, both female and minority students are being told (whether directly or indirectly) that serious study of mathematics and science is not for them. Thus, the system is not encouraging all those who might have an aptitude for and interest in studying engineering.

In contrast to most other professionals, future engineers (along with mathematicians and some scientists) tend to make their career choice in junior high/middle school. If they are not prepared and motivated to study engineering at that point, it is likely that they never will be.

Since the publication of *A Nation At Risk* more than a decade ago (U.S. Department of Education, 1983), it has been widely acknowledged that U.S. secondary school students have fallen behind their counterparts in most other industrialized nations in their knowledge of science and mathematics. Although average mathematics and science test scores in national assessments improved slightly during the 1980s, they are still well below those seen in the 1960s (National Science Board, 1991, p.14). Quantitative reasoning and problem-solving skills are particularly lacking, even in students who score well on standardized exams.

Inadequate mathematics and science preparation limits both the quality and the quantity of potential entrants to engineering. Many of those students who do enter engineering study are not prepared for its rigors, in terms of either knowledge or analytical skills. The result is students struggling to keep up, contributing to a high rate of attrition. In particular, inadequate preparation limits the participation of African Americans, Hispanic Americans, and other underrepresented minority groups, who lag their majority counterparts (and Asian Americans) in mathematics and science preparedness.

Over the past few years, many states have raised their standards for promotion and for high school graduation, revised teacher licensing and training practices, and improved the measurement of school performance. Other national reform efforts are being carried out. For example:

- The National Council of Teachers of Mathematics has established guidelines for mathematics curricula.
- The National Science Teachers Association is conducting a

study of science curricula and has completed a science curriculum guide for grades 6–12 (NSTA, 1993).

- The NSF has established both a Statewide Strategic Initiative (in 21 states) and an Urban Systemic Initiative (earmarked for the nation's 25 largest urban school districts) in an effort to transform the way U.S. schoolchildren learn about science, mathematics, and technology.
- The Division of Undergraduate Education of the NSF is managing Collaboratives for Excellence in Teacher Preparation, which bring together science and engineering faculty and education faculty to prepare future K–12 teachers.
- The National Research Council (NRC, 1989, 1990a, b) has issued several reports on mathematics curricula and teaching practices and has issued draft standards for K–12 science education (NRC, 1994), which will be released in 1995 as a companion to the mathematics standards.

Federal spending on precollege mathematics and science education has increased substantially in the past few years. According to "Special Tabulations" provided by the working group on the budget of the National Science and Technology Council Committee on Education and Training (estimate as of May 1994), the federal government is spending $955.431 million on science, mathematics, engineering, and technology education at the precollege level in fiscal year 1994. (This represents an increase of 85.7 percent over fiscal year 1991 spending; FCCSET, 1992.) In the White House, the National Science and Technology Council Committee on Education and Training coordinates these activities.

The main responsibility for improving the mathematics and science preparedness of students lies with the elementary and secondary schools. Together with parents, it is their responsibility to develop talent, encourage interest, and ensure that students persevere with math and science courses. Schools that fail to offer the necessary courses, or that eliminate potentially capable students by applying rigid criteria that do not allow for individual variation in abilities or background, restrict access unnecessarily. Teachers who are poorly prepared to communicate the attractions of science and engineering as careers also limit the potential talent pool. It is important for elementary and secondary school teachers to understand what engineering is (as distinct from science), so that they can advise and encourage potential engineering students.

"Student-to-student contact is particularly effective. Some ideas:

• Bring demonstrations to middle schools (e.g., a solar car team).
• Bring middle and high school students to campus, where college students can demonstrate equipment.
• Give college students credit for mentoring activities in working with middle/high school students."

G. Wayne Clough,
President,
Georgia Institute of Technology,
Personal communication to the BEEd,
February 28, 1994

However, higher education also has some responsibility for K–12 science, math, and "pre-engineering" education. Engineering schools cannot simply assume that an adequate supply of motivated, well-prepared students will always arrive at their doorstep. Direct outreach to K–12 students is vital to their mission. University faculty and laboratories may seem remote and abstract to precollege students and teachers alike. Direct contact is the best way to dispel that remoteness and impart a realistic understanding of what engineers and engineering students actually do. A few engineering schools are carrying on activities with precollege students—inviting them to visit, mentoring them, carrying design projects into K–12 schools as demonstrations, etc. However, such interactions are still uncommon.

> The distinction between science and technology needs to be addressed in K–12 and in universities. Our meekly accepting technological successes as "science achievements" and technological failures as "engineering catastrophes" is a direct result of our premise that engineering has its roots in science.
>
> David Kingery,
> Regents Professor,
> University of Arizona,
> Personal communication to the BEEd,
> March 3, 1994

Technological Literacy

Beyond the K–12 system, in this intensely technological era it is essential that as many members of society as possible understand the nature of technology, how it has transformed the modern world, and what are the contemporary issues involving engineering that are significant for the future of this culture, all of which make up a concept termed "technological literacy." This topic has important ramifications in that it affects the public support for engineering education and engineering endeavors, as well as having a strong impact on the number and quality of students interested in pursuing an engineering education.

In view of its educational mission, engineering education has a first-line responsibility for improving the technological literacy of the general public—especially for groups whose influence has direct impact on major political and economic decisions for society and on the engineering profession itself. One of the most effective routes to this goal is to increase the technological literacy of non-engineering students. To achieve this, it is necessary first to convince faculty throughout the university (including engineering faculty) of the importance of teaching non-engineering students about technology and their responsibility for doing this. Second, ways should be developed to do so economically and effectively using materials already developed and working with experienced, effective faculty both in engineering and in other fields. The materials developed in the New Liberal Arts program sponsored by the Alfred P. Sloan Foundation will serve as one good basis for this effort (Goldberg, 1990).

The BEEd believes that there are three components to technological literacy: knowledge of how objects and systems work, the social context within which engineering operates, and the cultural meaning of engineering. Because of the tendency for faculty to be narrowly focused, many engineering professors themselves know little about the broad field of engineering. This means that the teaching of technological literacy requires the new collegiality described above and will help bring faculty together to promote it.

Continuous Education of Engineers

Engineers in practice encounter two major types of change, namely, changes in the technological *content* of engineering knowledge and in the *context* of professional practice. Both circumstances tend to shorten the productive career lifetimes of engineers and thereby reduce the effectiveness of industry. The first type of change, in knowledge content, is predictable with an observable average period of about a decade in most fields. The second change, in practice context (such as economic and job stability, national goals, global trade patterns, etc.), is less predictable in period but is fairly rapid.

The challenge lies in the rapidity of change. Previously such change was on the time-scale of a career lifetime, whereas now and in the future many engineers will experience several change cycles over a career lifetime, each requiring the acquisition of new or updated knowledge (IEEE, 1995).

Given the large investment of educational resources and experience engineers represent, the nation cannot afford to view them as commodities, to be replaced when they become "obsolete." It is essential that engineering professionals continue to develop their knowledge and capabilities over a lifetime of practice. This will require a commitment to lifelong learning, which needs, in turn, the support of a continuing engineering education system and the motivation to use it.

"Refresher" courses, retraining, postbaccalaureate professional education, and continuing education are all viable means of minimizing the avoidable loss of engineers due to rapid technological obsolescence. Many private educational providers offer courses commercially, and the largest companies generally offer programs in-house. However, continuing education opportunities for engineers today are poorly integrated and not

PREACHING VS. PRACTICING

In a survey of industrial firms in surrounding states that was conducted by the University of Michigan College of Engineering (Atreya, 1994), 64 percent of surveyed companies said that they rank continuing education as either "high" or "medium" among their corporate priorities, but the same companies said that only about 30 percent of their professional and technical employees actually utilize continuing education opportunities. Incentives are not evident: only 5 percent of the responding companies require employees to earn continuing education credits; only 13 percent require employees to earn any other type of special certification; and 79 percent give no rewards or recognition for participation in continuing education activities. Significantly, 42 percent of the managers responding said that employees "lacked a sense of perceived need or payoff for participation."

"I believe that the central issue in continuing education today is a failure to communicate availability of services to the workforce and a failure to empower engineers to make decisions concerning their continuing education without seeking the approval of managers. In today's business climate of severe cost control, educational expenditures are often treated as "overhead" to be controlled, rather than as an investment which leads to improved return on investment.

Lionel V. Baldwin,
President,
National Technological University,
Personal communication to the BEEd,
November 24, 1993

A VIRTUAL UNIVERSITY

Carnegie Mellon University has proposed an experimental *Virtual University* that would encompass:

1. an interactive multimedia classroom where distance learners can participate real-time;
2. an on-line Internet library offering programs in digital video;
3. a portable classroom consisting of wireless equipment; and
4. interactive research facilities and research on interactive technologies.

This is only one example of many similar experiments now under way at U.S. engineering institutions. All such experiments recognize that other characteristics beyond these "virtual" ones are necessary to fulfill the overall educational mission of a university.

readily available to all engineers. For example, small and medium-sized companies generally lack the resources to mount effective training programs, and engineers in rural locations tend to have fewer opportunities than those located near cities.

Adult or continuing education overall is said to be a $30 billion per year business—larger even than the primary training enterprise. Nevertheless, continuing education and training of practicing engineers has never been a primary emphasis of the engineering education system. Although the opportunities for earning income through continuing education programs are substantial (especially since large U.S. corporations are reducing their in-house offerings in this area), only a few of the traditional baccalaureate institutions have pursued this opportunity aggressively. Engineering colleges typically offer short courses aimed at local industries. In some cases, these are televised or supplied as videotapes for viewing at industrial sites. However, the offering tends not to be broad. Content is usually geared to the research proclivities of individual faculty and may often be quite theoretical in nature. Marketing of these courses to the potential audiences is uneven. The incentives for attending (where they exist at all) are usually tied to company advancement rather than to the specific applicability of the knowledge gained.

No continuing education programs are subject to accreditation (nor does the BEEd call for that). Indeed, no standards currently exist for these offerings. The question is, then, by what means can the content and quality of these offerings be controlled, and how can their value be increased and their utilization expanded? Universities have a critical role to play.

It is vital to instill in engineering students both the skills needed to acquire continuous learning from various sources beyond the period of formal schooling and an understanding of the necessity for doing so. This will involve instilling an awareness of the sources of "distance learning" and exposing students to the mechanisms and techniques employed in accessing on-line instructional services of various kinds, both at home and at the work site. Concepts such as Carnegie Mellon's *Virtual University* (see box this page) and the Virtual

Online University[9] can be useful. Seminars presented by industrial representatives, such as adjunct professors, on their own experiences with continuous education and the value they have found in it could be quite effective as well.

[9]Established in 1994 by a group of educators who met on the Internet, Virtual Online University will begin its first term in spring 1995 with an initial on-line offering of 40–60 courses. Courses will meet regularly and include a combination of readings, exams, and research projects.

4

Achieving Change

STRUCTURAL ASPECTS AND ISSUES

In any social system (of which engineering education is one), organizational structure determines to a considerable extent the nature of both the processes (functions and actions) that can take place within the system and the products that result. Thus, structural features also have a large impact on the ability of the system to adapt to external and internal forces.

System Structure

It is worth examining the structure of the U.S. engineering education system briefly to ascertain its salient features and their implications for strategies to change the system so as to achieve the vision described in Chapter 2.

The nation's engineering education system includes not just higher education but also K–12, community colleges, and continuous (lifelong) engineering education. These elements are embedded in the larger U.S. society, whose political and economic influences typically affect engineering schools through the academic institution of which they are a part. Those socioeconomic and political factors also drive demand for engineers, as well as the supply, recruitment, and retention of engineering students.

In 1994, the system included 311 institutions that granted B.S. engineering degrees or higher in accredited programs. It incorporated not just the 150 or so "research universities" and "doctorate-granting" institutions but also the roughly 160 other institutions that focus primarily on undergraduate education and produce nearly a third of the nation's engineers.

Across these many institutions, there is great diversity in terms of size, age, traditions, research interests, departmental structure, strengths and weaknesses, and other characteristics. Some are urban; others are in rural locales. Some are small colleges within a comprehensive university; others are specialized technological institutions. Some are focused primarily on a specific engineering field (such as mining or chemicals, for example); and others are broadly balanced across fields.

The BEEd recognizes that the issue of scale is worthy of consideration here. How many schools and departments of engineering does the nation need to support? Is 311 accredited institutions the right number? Is it too many? Too few? In either case, how can the number be reduced or increased through external influence? Such questions are difficult, if not impossible, to answer. Yet they are questions of concern to academic administrators and to those in government (both federal and state) who have to find the funds to support engineering education. Realistically, in the U.S. system these determinations are made, however inefficiently, by the free market of supply and demand. Those market forces have produced the great diversity of engineering schools seen today, and no pronouncement by any external body—however authoritative—is likely to affect matters significantly.

The diversity of the nation's engineering education institutions is at once a great strength and a potential impediment to reform. Different characteristics imply differing needs and differing capabilities to change. One characteristic that most academic institutions share, however, is decentralized influence and authority at the level of the university, the department, and the individual. Academic freedom (and especially tenure) means, in effect, that each of these levels is relatively autonomous and thus is able to resist change. Consequently, it is difficult to impose major change within this system from the top down. A strong force in favor of stability is exerted by the Accreditation Board for Engineering and Technology[1] and the various regional accrediting bodies, which must review and in some cases approve changes in curriculum, degree requirements, etc.

Another characteristic of academic institutions is that they are vertically aligned organizationally. Vertical alignment means that a

[1]The Accreditation Board for Engineering and Technology is composed of 27 professional engineering societies; its accrediting business is carried out largely by volunteers from academe, industry, and government. The board establishes a "floor" of requirements for all engineering schools that wish to have their graduates considered as engineers and, consequently, want its endorsement. Above the floor are unlimited opportunities for schools to increase their quality and to exercise their unique missions.

school is separated organizationally and administratively from the rest of the university. It follows that collaboration between a school of engineering and a college of liberal arts, education, or business, for example, usually is difficult.

An important feature of the academic organizational structure is the role that the university as a whole plays in the creation of incentives in the engineering school. Although promotion and tenure recommendations are made at the department and college level, overall policy guidance is generated at the university level, and the final decisions on such matters are made by committees representing all academic units.

Finally, a major influence on the structure of universities is the fact that their funding is external. For public institutions, the state government determines some aspects of long-range policy through its support, and for all research universities, whether public or private, federal research funding has a powerful influence on the organizational structure and research/educational emphases of the institution.

These structural features tend to ensure that the overall system resists change. The walls between organizational units and the lack of autonomous ability to change direction, above the level of the individual, institutionalize a structural rigidity and conservatism.

Implications for Change Strategies

The most obvious implication of the structural rigidity inherent in the engineering education system is that change must be effected at the "local" level—that is, at the level of the school, department, or individual. For a variety of reasons (see Massy et al., 1994), it is already difficult to achieve consensus on needed changes even at the departmental level. The more elements of the system that must be engaged, the more difficult the change will be to effect.

Certainly a significant factor affecting the potential for change, however, is the imposing workload that engineering faculty face daily. Although individuals may be, in principle, more amenable to change than other levels of the system, it is generally difficult for them to respond to additional demands on their time—demands that any form of change usually imposes.

Thus, it is impossible to be prescriptive about actions that should be taken. Both in substance and in process, any modifications must be adapted to local values and circumstances and must recognize the pressure they place on already stressed individuals and organizations. The diversity of institutions makes it likely that a "free market" approach to change will be more effective than any central mandate. Such changes will require the cooperation of the Accreditation Board for Engineering and Technology and the regional accrediting organizations; their close involvement in this process on a national level is essential.

Another implication of the structural characteristics of the system is that collaboration across organizational and disciplinary barriers must be emphasized. The effort required to break down barriers may be large, but it can be anticipated that the benefits might also be unexpectedly large.

The most powerful change agent for many of the 311 engineering institutions, however, may be federal agency funding policy. Federal funding is what created the present research-oriented structure of academic engineering in the first place, and it can bring about change faster than any other influence—especially among the research universities. Indeed, this process of "cultural change" is already well under way through programs such as NSF's Engineering Research Centers, Engineering Education Coalitions, and alliances for minority participation and the manufacturing education and training awards of the multi-agency Technology Reinvestment Project.

STRATEGY FOR CHANGE

The task of the BEEd has been to:

- understand the external forces impinging on engineering education and driving the need for change (Chapter 2);
- formulate a vision of the future of engineering education (Chapter 2);
- assess the current state of engineering education and identify the challenges it faces in realizing that vision (Chapter 3);
- understand how the structure of the engineering education system affects the possibilities for achieving needed change (Chapter 4); and
- develop the outlines of a plan to achieve needed changes.

That plan follows in Chapter 5. The BEEd wishes to emphasize, however, that this study represents a preliminary effort, the results of which are necessarily generalized and qualitative. The board has provided an overview—a top-level analysis. It remains for the engineering education community at large to perform the follow-on work in the context of their local circumstances and to make the detailed changes needed to achieve the vision in terms that make sense within their particular institutional setting.

Thus, in Chapter 5 the BEEd issues a call to action for all those who have a stake in the performance of the engineering education system and the quality of its products.

5

A Call to Action

The BEEd believes that the time for discrete, disconnected reactions to the forces and conditions shaping engineering education is past. *The BEEd calls for engineering educators to work together nationally to improve the engineering education system.* That may mean redesigning some aspects of the system. It will mean improving the function of each element of the system and integrating it—that is, bringing the various elements into better balance with each other. Perhaps most important, it will mean building into engineering education—especially at the institutional level—the capacity to adapt flexibly to rapid and continuous changes in technology (both industrial and educational) in the economy (global, national, and regional), in student demographics, in industrial demand, and in national priorities. Indeed, some of those changes will be brought about through changes in engineering education itself.

If substantial and necessary change is to occur throughout the engineering education system, performance evaluation and feedback are needed on the extent to which new engineers meet the needs of society in the twenty-first century—that is, on how well they satisfy the vision outlined in Chapter 2 (and summarized in Table 1-1). Feedback also is needed on the extent to which engineering education is meeting the needs of the practicing engineers in the twenty-first century.

Evaluation of the characteristics and responsiveness of the engineering education system itself is needed in order to create a more flexible system capable of responding effectively to changing needs and circumstances. The system, at the level of individual schools and departments of engineering, must have a strong capacity for self-evaluation and adaptation and a willingness to undergo such changes. The effectiveness of the system will hinge on the willingness of every institution and every faculty member to listen; to be aware; to shape as well as respond to change; and to alter their collective outlook, programs, and approaches accordingly. It will also depend on the ability of the engineering accrediting organizations (the Accreditation Board for Engineering and Technology in particular) to develop measurable performance- or output-oriented accreditation criteria that encourage such changes. Engineering educators must strive for flexibility and adaptability in everything they do as educators.

Therefore, given the decentralized and diversified nature of the engineering educational system, it is essential for each engineering institution to update itself within the context of an institutionally shared vision of the overall system and its goals—a concept best expressed by the phrase "think globally, act locally." Such an undertaking will involve certain actions common to all schools; other actions will depend on the specific character and mission of the individual school and will be identified through self-assessment and collective discussion of the institution's goals and areas requiring change.

Engineering schools are the core units of the system, but there are also essential actions that must be taken by industry, government, accreditation bodies, and the professional societies, either alone or in conjunction with the academic and other sectors. Attaining the vision will require coordinated action across the entire system.

ACTIONS FOR *ALL* INSTITUTIONS

Conduct Institutional Self-assessment

As a first step, each engineering institution (at the school and department level) should undergo a process of self-assessment and self-evaluation from the standpoint of the vision and goals enunciated in this report. This process should be a collegial one involving participation that is as broad as possible among administrators, faculty at all levels (including faculty from other schools or colleges in the institution), selected students, alumni and alumnae, and major employers of the graduates. An effort should be made to define a "profile" of the institution and its characteristics and then to discuss the actions described below in the context of that profile. Ideally, the output of this self-assessment should be a consensus document in the form of a strategic plan or the equivalent, which is published and circulated within the institution. For an example of such a document, see

Massachusetts Institute of Technology's *Long Range Plan for the School of Engineering, 1994–1998* (MIT, 1994).

Subsequently, administrators and faculty throughout the institution should *monitor on a continuing basis the implementation of the action plan* vis-a-vis the specific elements of the vision and call to action presented in this report. Feedback from industry and from graduates (the "customers" of the engineering education enterprise) should be solicited as input to that monitoring function.

Redress Imbalances in the Faculty Incentive System

Following the institutional self-assessment, it may be anticipated that one of the highest priorities will be to redress imbalances in the faculty incentive system, particularly in research universities.[1] This will likely entail the following:

Align the faculty reward system more fully with the total mission and purpose of the institution. The reward system at each institution across the existing system must ensure a proper balance among teaching, research, service, and professional activities. In assessing intellectual attainment and creativity of faculty, Ernest Boyer, in *Scholarship Reconsidered* (Boyer, 1991), urges that the quality of scholarship be assessed over four areas of activity: (1) scholarship of discovery (commonly called research), (2) scholarship of integration (synthesis within and across disciplines), (3) scholarship of application (professional use of knowledge), and (4) scholarship of teaching (transformation and communication of knowledge). Institutions should examine their promotion and tenure policies to ensure that appropriate weight is given to each area in which there is documented evidence of achievement and that the activity of faculty is balanced over a career path.

Expand the working definition of scholarship to include "pedagogy" (research and development on teaching methods and curriculum development), and redefine "publications" to include formal curriculum model development, multimedia teaching approaches, and the creation of tutorial modules.

Expand the definition of creative research activity to incorporate measures of industrial relevance (e.g., technology transfer) in assessing faculty performance.

Develop and/or monitor and adopt criteria and practices for the evaluation of teaching effectiveness. The BEEd notes, as an example,

[1]While the actions described under this heading are directed toward research universities and other doctoral-degree-granting institutions, there is a clear trend among comprehensive universities to place more emphasis on research as well.

the interesting pilot project being conducted by the American Association for Higher Education on peer review of teaching (AAHE, 1994). The methods must, however, avoid attempts to treat evaluation too quantitatively, as merely the results of written testing programs or numerical student evaluations.[2]

Improve Teaching Methods and Practices

Improve teaching methods by exploring and experimenting with such techniques as active (participatory) learning, expanded use of educational technology, increased faculty awareness of cognitive science findings, cooperative learning, peer teaching, team teaching, case studies, and "competency-based" (i.e., involving demonstration of integrated skill and knowledge) assessment of students' knowledge. Ways could be found to reward faculty for successful implementation of alternative styles of delivery.

Provide training (or access to training) in teaching skills for academic-track Ph.D.s and faculty recruits, as well as refresher training for more senior faculty. For foreign-born teaching assistants and faculty, this could include language and cultural "sensitivity" training.

Find creative ways to utilize more engineers from industry in teaching, especially teaching of undergraduates.

Ensure greater participation by faculty (as opposed to teaching assistants) in teaching undergraduates, and emphasize student-faculty interaction. For imparting motivation and "connectedness" to the educational experience, nothing can replace direct personal contact with a respected faculty mentor—a fact that must be kept in mind as "distance learning" receives greater emphasis.

Strive to create a positive, supportive climate for engineering students by emphasizing success and personal encouragement rather than the "weeding-out" approach that has often been taken in the past.

Establish mechanisms to provide faculty members with greater exposure to engineering practice, such as:

- recognizing relevant types of consulting in promotion and tenure evaluations;
- providing industrial sabbaticals;
- encouraging joint research with industry colleagues and adjunct faculty; and

[2]More useful than the standard student evaluation of teachers would be evaluations based on surveys of students five years beyond graduation who are actively involved in their careers. Such evaluations would likely be more objective and reliable than those made by current students.

- recognizing the study of engineering practice as bona fide research.

"It is time for the four-year engineering degree to join the slide rule, log tables, the French curve, and ammonia-reeking blueprints as artifacts of the past."
(Augustine, 1994b)

Ensure That the Curriculum Supports the Institution's Strategic Plan

BEEd members share the growing recognition that four years is no longer enough time for the formal education of an engineer about to enter professional practice. *Schools must consider and implement, as appropriate, alternative paths to the undergraduate degree*, including:

- a "general engineering" degree;
- three- or four-year pre-engineering programs leading into a graduate engineering degree program;
- a cooperative (i.e., work-study) degree; and
- a five-year bachelor's degree.

Again, these are not prescriptions but suggestions to be considered in the context of each institution's local circumstances.

Consider and implement, as appropriate, alternative paths to graduate degrees, including:

- a practice-oriented master's degree;
- a combined bachelor's/master's degree;
- a Ph.D. or D.Eng. with an industrial research and development track; and
- a practice-oriented doctorate.

Any reforms of graduate engineering education should be addressed as integral parts of the combined B.S./M.S. and graduate-track pre-engineering programs.

Develop practice-oriented graduate study modules aimed at engineers in practice. Such modules could be developed by joint industry/faculty teams. They might consist of two, three, or four courses and would be aimed at meeting contemporary practice or research needs. They would not result in a graduate degree but could be credited toward such degrees, if pursued, at a later time.

Pursue appropriate undergraduate curricular reform, including the following, for example:

- Ensure early exposure to engineering practice and a sense of the role of engineers in society.

- Provide for more-extensive exposure to hands-on, industrial practice aspects, team work, and creative design.
- Ensure that the curriculum reflects current and emerging technology and tools (e.g., modeling and simulation, finite element analysis, risk analysis).
- Emphasize interdisciplinary education and "systems" thinking.
- Monitor ongoing experiments in curricular reform (e.g., engineering education coalitions), and implement aspects pertinent to the institution, ensuring continued strong grounding in engineering science and math.
- Include requirements for coursework in business and in programs dealing with science, technology, and society (or the equivalent), emphasizing sustainable development of the environment.
- Review non-engineering course requirements, including liberal arts, with a view toward improving the communication skills of engineers; broadening their horizons; and preparing them to be more effective professionals, citizens, and leaders.
- Experiment with ways to inject into the curriculum some exposure to the international aspects of industrial competition and technology development (e.g., student exchanges, seminars by foreign and industry adjunct faculty, use of foreign examples in teaching).
- Instill in students a desire for continuous and lifelong learning to promote professional achievement and personal enrichment.

Expand Beneficial Interactions and Outreach

Pursue diversity of the student body by:

- improving access, that is, ensuring that all who could benefit from an engineering education are prepared to obtain and *can* obtain one;
- conducting an institutional self-assessment of the diversity of the student body and campus climate to identify needed corrective actions;
- taking steps to create a positive, supportive climate that ensures racial, gender, and ethnic diversity among all engineering students (including the creation of advisory services);
- establishing formal commitments and incentives to bring the demographic diversity of the faculty and student body into balance with each other; and
- improving the articulation with community colleges and providers of continuing education.

The pursuit of diversity must be accompanied by a continued commitment to excellence in engineering education.

Improve faculty diversity. Institutions should move toward the following self-imposed goals, with the recognition that the appropriate mix of faculty characteristics will differ for institutions of different mission:

- Achieve greater diversity of race, gender, ethnic background, and age by altering the mix of faculty characteristics through self-initiated actions by department heads, deans, and university administrators; and
- Employ on the faculty more engineers from private industry and government who have engineering design experience and management experience and who have demonstrated good teaching abilities.

Strive to develop a "new collegiality"—a shared sense of mission and purpose that will better integrate both the faculty and the process of engineering education. An effective way to do this is for the faculty to collaborate in developing lower-division courses in engineering. Undertaking the self-assessment called for earlier also will promote collegiality.

Establish/improve coordination with the rest of the university—for example, to consider holistically the undergraduate program of all students. One goal should be to ensure that non-engineering undergraduates obtain a better understanding of engineering and technology through one or more survey courses.

OTHER POSSIBLE ACTIONS FOR CONSIDERATION

Actions to be Undertaken by Institutions

Each engineering education institution must identify and undertake actions necessary to update its practices and outlook in accordance with the vision described earlier and with its own strategic plan developed through self-assessment and self-evaluation. Following are examples of actions that may be appropriate for some schools. The BEEd emphasizes that this list is by no means all-inclusive.

Specialize the institution's program offerings to focus available resources, building on established strengths to maintain excellence and maximize cost-effectiveness in those areas. Forthright strategic planning will be needed. Collaboration with other academic institutions for dividing up responsibilities or for sharing of equipment and facilities will grow more important. Collaborations with private industry and government agencies will help to identify the optimal "profile" of each institution (PCAST, 1992).

Consider alternatives to tenure such as fixed-year contracts for all faculty.

Provide time for faculty professional development, emphasizing participation in collaborative curriculum development efforts and major industrial and government research projects of a cross-disciplinary nature. Give credit for active participation in professional societies.

Document excellent teaching. Develop written profiles of exemplary teachers and case studies of successful experiments with innovative teaching methods.

Especially for schools in urban or industrialized areas, *develop cadres of part-time faculty from among practitioners*, or establish a new track of "co-op faculty" slots that would be filled full-time on a revolving basis, perhaps a quarter at a time, by industry practitioners. Employ early retirees from industry as regular full-time faculty, especially in manufacturing and design areas. (It should be noted, however, that not all industry professionals will be good teachers.)

Develop curricular models (and instructional modules) from interdisciplinary building-blocks—perhaps in collaboration with other engineering schools, and consider the possibility of "modularizing" the curriculum for greater flexibility.

Become more international in orientation and programs. Respond to state or regional efforts to increase foreign trade and to the needs of the often large contingent of foreign students.

Formally recognize the pursuit of technological literacy among the general population as part of the school's mission. To that end, actions might include the following:

- Require *all* non-engineering undergraduates in the institution, including science and math education majors, to take one or two survey courses on engineering and technology.
- Establish, through statewide consortia, centers where K–12 teachers could acquire in-service training on teaching tools and topics in support of technological literacy.
- Conduct a pre-service "summer school" for college students majoring in science or math education.
- Establish mechanisms by which some engineering graduates would teach K–12; one route might be as visiting ("per diem") teachers in exchange for accelerated acquisition of a teaching certificate.
- To the extent possible, involve parents in the K–12 programs.

Recognize the institution's responsibility for improving K–12 science, math, and "pre-engineering" education, especially with a view to quantitative reasoning and problem-solving skills. Actions might include the following:

- Provide on-campus tutorials for K–12 teachers in the effective use of computer-based learning technologies.
- Provide "packaged" laboratory projects that can be used in the K–12 classroom.
- Encourage engineering faculty to establish partnerships with K–12 teachers.
- Encourage faculty to establish mentoring relationships with middle- and high-school teachers and students, perhaps utilizing electronic networking.

Action to be Undertaken by Industry

The BEEd urges companies to consider the following possible actions:

Remove barriers and provide incentives to engineers to pursue continuing technical education.

Adopt a sabbatical system to reward employees with continuing education options, and encourage them to pursue these options without fear of adverse career implications.

Change the corporate reward structure to accommodate releasing professionals to teach in universities for a limited period of time.

Encourage engineering staff to participate in engineering education development activities such as those conducted by the Accreditation Board for Engineering and Technology, American Society for Engineering Education, and engineering school advisory boards.

Fund fellowship programs and scholarships for women and minority engineering students.

Make available a larger number and range of summer internships, particularly for undergraduates.

Fund faculty fellowships, internships, and adjunct professorships.

Provide engineering instructional materials to K–12 schools, and encourage professionals to partner with K–12 teachers in providing hands-on engineering experiences for students.

On their own initiative, successful graduates should contact their professors and departments with an offer to *speak to engineering student groups regarding their personal career experience in industry.*

Actions to be Undertaken by Professional Societies

The Engineering Deans' Council or other appropriate group should *continue working cooperatively with the Accreditation Board for Engineering and Technology in its reassessment of accreditation criteria* in accordance with the types of changes suggested in this report and implemented in response to current and future needs in engineering education.

In addition to rewarding excellence in research, *societies should place emphasis on honoring faculty excellence in education.*

Although it is recognized that the societies compete, to some extent, with universities and other providers of continuing engineering education, this is nevertheless an obvious area in which societies can *collaborate with universities*, to the mutual advantage of all participants.

The societies should consider *holding more education sessions at technical conferences.*

Engineering societies can do much to *assist universities in recruiting engineering students,* especially through effective information dissemination about the nature and appeal of engineering. (Engineers' Week, held in February of each year, is an excellent example.)

Engineering societies can *encourage their members to partner with K–12 teachers in providing hands-on engineering experiences to students.*

Actions to be Undertaken by Government

NSF should take steps to *disseminate and implement the results of the engineering education coalitions on a systemwide, evolutionary basis* as they become available. Resulting curriculum modification and application efforts at various institutions should be monitored and reported on a nationwide basis, perhaps through the National Engineering Education Delivery System, as it becomes established.

NSF could *expand its existing Course and Curriculum Development program, which works to develop teaching tools for use by engineering educators.*

NSF could *fund U.S. faculty members to review foreign emerging technology in their field and report in published papers and lectures.*

Actions to be Undertaken by Government-Industry-University Cooperatives

A coalition of university and industrial organizations, with federal coordination and funding, should *develop multimedia network(s) on which continuing education courses can be made more widely avail-*

able on live/interactive television or on videotape. The National Engineering Education Delivery System is one such network that should be widely supported.

Develop a nationwide instructional television network for undergraduate instruction. The model for this concept is the National Technological University, which is directed at practicing engineers. Such a network would be an expanded version of regional or intercampus television networks now in place in Utah, North Carolina, Colorado, and elsewhere. A variety of interinstitutional issues such as copyright and compensation would have to be resolved.

Establish an on-line electronic library of documents that contains a number of one-on-one tutorials, or "learning modules," for use by engineers and students. This "living electronic handbook" should be made available through the National Engineering Education Delivery System.

Actions to be Undertaken by the Accrediting Authority

The Accreditation Board for Engineering and Technology should *adopt, whenever possible, measurable performance- or output-oriented accreditation criteria for engineering programs.* This means, in general, placing greater emphasis on the quality of graduates and of research than on inputs (i.e., the number of students and faculty and the amount of financial support for research and education).

Actions to be Undertaken by Other Groups of the Engineering Community

Representatives of the engineering community, perhaps convened through the Engineering Deans' Council, should *explore educational innovations, initiatives, and practices in other countries* that appear to be effective in producing high-quality engineers, and should report these widely.

The engineering education community, perhaps through the National Research Council, should *proactively support ongoing efforts to reform K–12 science and mathematics* at the national, state, and local levels.

A task force should be established, perhaps through the National Research Council, to *examine the college curricula of education students who are planning to teach K–12 math and science from the standpoint of technological literacy and the presentation of engineering awareness and examples of engineering achievements.* An effort should be made to "re-invent" many undergraduate science courses for K–6 teachers. The task force might even design one or more textbooks to introduce engineering and technology to K–12 students.

EPILOGUE

The BEEd is well aware that major changes in large, decentralized systems such as the engineering education system seldom take place in direct response to a single stimulus such as this report. Rather, such changes usually reflect a gradual shifting of opinions, attitudes, and practices arising from a recognition and clearer understanding of new external conditions and concomitant new internal needs and emphases. The Beed believes that such a change is occurring in engineering education—indeed, at some places in the system it is well under way, and recently a number of authoritative reports have urged changes similar to those that the board recommends. The BEEd hopes that a special contribution of this report, based on discussions with a very broad cross-section of the engineering community, will be to provide a clearer view of the specific areas where change is needed and to suggest workable mechanisms for achieving positive change.

However, the BEEd's work necessarily has had a finite scope and duration. The real work of implementing needed changes in the engineering education system is both the individual and collective responsibility of the multiple constituencies whose concern is engineering education: academic administrators and faculty members, government policy makers and agency program managers, professional society leaders, and industrial leaders. The work must continue over time. It will require a sustained commitment, together with self-assessment and the will to continue adapting to new circumstances. The education of this nation's engineers deserves no less.

References

AAES. 1993. Engineering Degree Trends, 1993. Engineering Workforce Bulletin No. 128. Washington, D.C.: American Association of Engineering Societies.

AAES. 1994. Engineering Enrollment Highlights, Fall 1993. Engineering Workforce Bulletin No. 132. Washington, D.C.: American Association of Engineering Societies.

AAHE. 1994. 12 Universities to Pilot New Methods for Evaluating Teaching. Press release, 26 January 1994. Washington, D.C.: American Association for Higher Education.

ASEE. 1955. Report of the Committee on Evaluation of Engineering Education. Journal of Engineering Education 46:26–60.

ASEE. 1994. Engineering Education for a Changing World: Project Report. Washington, D.C.: American Society for Engineering Education.

Atreya, E. 1994. Continuing Engineering Education Needs: 1993 Survey of Managers. Report to Engineering Conference, College of Engineering, University of Michigan, Ann Arbor, Michigan.

Augustine, N.R. 1994. Engineering the Next Century. Address to the Annual Meeting of the National Academy of Engineering, 5 October 1994, Washington, D.C.

Boyer, E.L. 1991. Scholarship Reconsidered: Priorities of the Professorate. Lawrenceville, New Jersey: Princeton University Press.

Campbell, G., Jr. 1992. The gender gap in minority engineering education. National Action Council for Minorities in Engineering Research Letter 3(1):1–7.

Carmichael, J.W., Jr., and J.P. Sevenair. 1991. Preparing minorities for science careers. Issues in Science and Technology 7(3):55–60.

Dertouzos, M.L., R.K. Lester, R.M. Solow, and The MIT Commission on Industrial Productivity. 1989. Made in America: Regaining the Productive Edge. Cambridge, Massachusetts: The MIT Press.

Drexel University. 1992. E4: Enhanced Educational Experience for Engineers—Summary Report. Philadelphia, Pennsylvania: Drexel University Press.

Drexel University. 1994. Retention rate of E4 students in all cohorts. Internal memorandum by J.H. Gregory, 23 August 1994, Drexel University.

FCCSET. 1992. By the Year 2000: First in the World. Report of the FCCSET Committee on Education and Human Resources. FY 1993 Budget Summary. Washington, D.C.: Office of Science and Technology Policy.

Florman, S.C. 1987. The Civilized Engineer. New York: St. Martin's Press.

Goldberg, S., ed. 1990. The New Liberal Arts Program: A 1990 Report. New York: Alfred P. Sloan Foundation.

IEEE. 1995. Industry 2000: Technical Vitality Through Continuing Education. Proceedings of the meeting of the IEEE Educational Activities Board, Denver, Colorado, 12–13 May 1994. Piscataway, New Jersey: Institute of Electrical and Electronics Engineers, Inc.

Kapoor, S. 1994. Lessons learned from interdisciplinary approaches in manufacturing education. Paper presented at the National Science Foundation Workshop on Restructuring Engineering Education: A Systems Approach, 6–9 June 1994, Washington, D.C.

Kerr, C. 1994. Knowledge ethics and the new academic culture. Change 26(1):8–15.

Massy, W.F. A.K. Wilger, and C. Colbeck. 1994. Overcoming "hollowed" collegiality. Change 26:11–20.

McClintock, F.A. 1994. Learning Modules for Engineering, Both In-School and On-the-Job (LeMode). Draft prospectus, Department of Mechanical Engineering, Massachusetts Institute of Technology.

McMasters, J.H., and B.J. White. 1994. Report on the Second Boeing-University Workshop on: An Industry Role in Enhancing Engineering Education, 18–19 July 1994, Seattle, Washington: The Boeing Company.

MIT. 1994. Long-range plan for the School of Engineering, 1994-1998. Cambridge, Mass.: Massachusetts Institute of Technology.

NAE. 1991. Engineering as a Social Enterprise. National Academy of Engineering, National Research Council. Washington, D.C.: National Academy Press.

National Science Board. 1991. Science and Technology Indicators, Tenth Edition (NSB 91-1). Washington, D.C.: National Science Foundation.

National Science Board. 1993. Science and Engineering Indicators—1993 (NSB 93-1). Washington, D.C.: National Science Foundation.

NRC. 1985. Engineering Education and Practice in the United States: Foundations of Our Techno-Economic Future. Committee on the Education and Utilization of the Engineer, National Research Council. Washington, D.C.: National Academy Press.

NRC. 1988. Foreign and Foreign-Born Engineers in the United States: Infusing Talent, Raising Issues. Committee on the International Exchange and Movement of Engineers, Office of Scientific and Engineering Personnel, National Research Council. Washington, D.C.: National Academy Press.

NRC. 1989. Everybody Counts: A Report to the Nation on the Future of Mathematics Education. Committee on the Mathematical Sciences in the Year 2000, Board on Mathematical Sciences and Mathematical Science Education Board, National Research Council. Washington, D.C.: National Academy Press.

NRC. 1990a. Reshaping School Mathematics: A Philosophy and Framework for Curriculum. Mathematical Sciences Education Board, National Research Council. Washington, D.C.: National Academy Press.

NRC. 1990b. On the Shoulders of Giants: New Approaches to Numeracy. Mathematical Sciences Education Board, National Research Council. Washington, D.C.: National Academy Press.

NRC. 1993. Major Issues in Engineering Education: A Working Paper of the Board on Engineering Education. Board on Engineering Education, National Research Council. Washington, D.C.: National Academy Press.

NRC. 1994. National Science Education Standards—Draft (1 December 1994). National Committee on Science Education Standards and Assessment, National Research Council, Washington, D.C.

NSF. 1988. News release: NSF Funds Undergraduate Engineering Curriculum Development (NSF PR 88-52). Washington, D.C.: National Science Foundation.

NSF. 1992. Science and Engineering Degrees, by Race/Ethnicity of Recipients: 1977–90. Surveys of Science Resources Series (NSF 92-327). Washington, D.C.: National Science Foundation.

NSF. 1993. Program Announcement: Engineering Education Coalitions. Washington, D.C.: National Science Foundation.

NSF. 1995. Restructuring Engineering Education: A Focus on Change (NSF 95-65). Arlington, Virginia: National Science Foundation.

NSTA. 1993. The Content Core: A Guide for Curriculum Designers (Vol. 1). Arlington, Virginia: National Science Teachers Association.

PCAST. 1992. Renewing the Promise: Research-Intensive Universities and the Nation. Report of the President's Council of Advisors on Science and Technology. Office of Science and Technology Policy, Executive Office of the President. Washington, D.C.: U.S. Government Printing Office.

Ramo, S. 1988. The Business of Science: Winning and Losing in the High-Tech Age. New York: Hill and Wang.

SPEE. 1930. Report of the Investigation of Engineering Education, 1923-1929. Urbana, Illinois: Society for the Promotion of Engineering Education.

SPEE. 1940. Aims and scope of engineering curricula. Journal of Engineering Education 30(7):555–566.

SPEE. 1944. Report of the Committee on Engineering Education After the War. Journal of Engineering Education 34(9):589.

Seymour, E., and N.M. Hewitt. 1994. Talking About Leaving: Factors Contributing to High Attrition Rates Among Science, Mathematics, and Engineering Undergraduate Majors. Final report to the Alfred P. Sloan Foundation on an ethnographic inquiry at seven institutions. Boulder, Colorado: Bureau of Sociological Research, University of Colorado.

U.S. Department of Education. 1983. A Nation at Risk: An Open Letter to the American Public. A Report of the National Commission on Excellence in Education. Washington, D.C.: U.S. Department of Education.

Vetter, B.M. 1992. Diversity in the engineering environment. Pp. 91–96 in Proceedings of the Women in Engineering Conference: Increasing Enrollment and Retention, Washington, D.C., 31 May–2 June, 1992. West Lafayette, Indiana: Women in Engineering Program Advocates Network.

White, R.M. 1991. Science, engineering, and the sorcerer's apprentice. The Bridge 21(1):13–20.

Appendix A

BEEd Task Statement

ENGINEERING EDUCATION: DESIGNING AN ADAPTIVE SYSTEM

Following a series of regional symposia ending March 1994 at which "*Major Issues in Engineering Education: A Working Paper of the Board on Engineering Education*" will have been presented and discussed, the Board plans to write a report on the status of engineering education. The Board, in its new report, will characterize, analyze, and rank by importance the issues and challenges, present the Board's conclusions, and provide recommendations appropriate for action by the engineering education system of the future which will encompass a student population that will be diverse in ethnicity and gender. The report will present a vision of what engineering education should be in the 21st century and will address the following: supply issues in the student pipeline with emphasis on women and other groups now underrepresented in engineering; the undergraduate experience, including curriculum; graduate education, including the need to specialize to focus resources, practice-oriented masters and doctoral degrees, and participation of foreign nationals; the state of the engineering professoriate; continuing education of engineers and other technical personnel; the cost of an engineering education; and technological literacy.

Appendix B

List of Regional Symposia and Particpants

MAJOR ISSUES IN ENGINEERING EDUCATION: A WORKING PAPER

First Regional Symposium
November 12, 1993
Chicago, Illinois
Total Participants 32 (21 Guests, 8 BEEd Members, 3 NRC Staff)

Second Regional Symposium
January 24, 1994
Dallas/Ft. Worth, Texas
Total 50 Participants (37 Guests, 10 BEEd Members, 3 NRC Staff)

Third Regional Symposium
February 28, 1994
San Francisco, California
Total 54 Participants (39 Guests, 11 BEEd Members, 4 NRC Staff)

Fourth Regional Symposium
March 24, 1994
Washington, D.C.
Total 107 Participants (89 Guests, 13 BEEd Members, 5 NRC Staff)

Bruce Alberts
President, National Academy of Sciences

M. Dayne Aldridge
Director, Thomas Walter Center for Technology Management, Auburn University

Betsy Ancker-Johnson
Vice President, General Motors Corporation (retired)
Chair, World Environment Center

Alfredo H-S. Ang
Professor of Civil Engineering, University of California, Irvine

Bruce W. Arden
Dean, Engineering Department, University of Rochester

William Howard Arnold
President, Louisiana Energy Services

Lionel V. Baldwin
President, National Technological University

Robert F. Barfield
Dean, College of Engineering, University of Alabama in Tuscaloosa

Philip Barkanz
Professor of Mechanical Engineering, Design Division, Stanford University

Joel W. Barlow
Professor, Department of Chemical Engineering, University of Texas at Austin

Eleanor Baum
Dean of Engineering, The Cooper Union

Martin Becker
Dean of Engineering, University of Miami

Louis A. Beecherl, Jr.
Member, Engineering Foundation Advisory Council, College of Engineering, University of Texas at Austin

David P. Billington
Professor of Civil Engineering and Operations Research, School of Engineering and Applied Science, Princeton University

W. Murray Black
Interim Associate Dean for Research and Graduate Studies, School of Information Technology and Engineering, George Mason University

John G. Bollinger
Dean, College of Engineering, University of Wisconsin–Madison

Arthur J. Bond
Dean of Engineering and Technology, School of Engineering & Technology, Alabama A&M University

Joseph Bordogna
Assistant Director for Engineering, National Science Foundation

Harry E. Bovay, Jr.
President, Mid-South Telecommunications Company

Sidney A. Bowhill
Professor of Electrical Engineering, University of Lowell

George Bugliarello
President, Polytechnic University

Peter Cannon
Vice Chair, Board on Engineering Education
Managing Partner, V.R.E.

Stephen H. Carr
Associate Dean, Undergraduate Engineering, Robert R.McCormick School of Engineering and Applied Science, Northwestern University

Ben H. Caudle
B.J. Lancaster Professor of Petroleum Engineering, Department of Petroleum Engineering, University of Texas at Austin

Claude Cavender
Lt. Col. U.S. Air Force Ret.

David C. Chang
Dean of Engineering, College of Engineering & Applied Sciences, Arizona State University

Blake Cherrington
Dean, Engineering and Computer Science Department, University of Texas at Dallas

Robert P. Clagett
Executive Director, Rhode Island Technology Transfer Center (RITTC)

G. Wayne Clough
Provost, University of Washington

Roy H. Cornely
Professor, Department of Electrical and Computer Engineering, New Jersey Institute of Technology

Ross B. Corotis
Associate Dean, Hackerman Professor of Civil Engineering, Whiting School of Engineering, The Johns Hopkins University

Harvey G. Cragon
Professor, Department of Electrical and Computer Engineering, University of Texas at Austin

Jose B. Cruz, Jr.
Dean of Engineering, College of Engineering, Ohio State University

Edward L. Cussler
President, American Institute of Chemical Engineers

John Daily
Chair, Mechanical Engineering, College of Engineering and Applied Science, University of Colorado

James W. Dally
Professor, Mechanical Engineering Department, College of Engineering, University of Maryland

Daniel B. DeBra
Professor of Aeronautics and Astronautics, Stanford University

Eugene M. Deloatch
Dean, School of Engineering, Morgan State University

Anthony J. DeMaria
Assistant Director of Research, United Technologies Research Center

Avery H. Demond
Assistant Professor, Environmental and Water Resources Engineering, University of Michigan

Denice D. Denton
Associate Professor, Department of Electrical and Computer Engineering, University of Wisconsin–Madison

Charles A. Desoer
Professor Emeritus, Department of Electrical Engineering and Computer Sciences, University of California, Berkeley

Bradley W. Dickinson
Associate Dean, School of Engineering and Applied Science, Princeton University

George E. Dieter
Dean of Engineering, College of Engineering, University of Maryland College Park

Frederick H. Dill
Research Staff Member, IBM, Thomas J. Watson Research Center

Francis G. Dwyer
Senior Scientist and Manager, Catalyst Research and Development, Mobil Research and Development Corporation

James Economy
Professor and Head, Department of Materials Science and Engineering, University of Illinois

A.J. Edmondson
Department of Mechanical & Aerospace Engineering, University of Tennessee at Knoxville

Roger Eichhorn
Dean of Engineering, University of Houston

Carl Erdman
Executive Associate Dean of Engineering, Project Director, Foundation Coalition, Texas A&M University

Edward W. Ernst
Allied Signal Professor of Engineering, School of Engineering, University of South Carolina

H. Chik M. Erzurumlu
Dean of Engineering, Portland State University

D. L. Evans,
Professor and Director, Innovation in Engineering Education Program, College of Engineering and Applied Sciences, Arizona State University

Richard Evans
Graduate Assistant and Doctoral Student, School of Information Technology and Engineering, George Mason University

David Farrell
Natural Resources and Systems, Agricultural Research Service, U.S. Department of Agriculture

Samuel C. Florman
Vice President, Kreisler Borg Florman Construction Company

Norman L. Fortenberry
Program Director, Engineering, Division of Undergraduate Education, National Science Foundation

Karen Frair
Associate Dean of Engineering, College of Engineering, University of Alabama

James Freeman
College of Engineering, San Jose State University

David L. Freyberg
Associate Professor, Department of Civil Engineering, Stanford University

Robert B. Fridley
Executive Associate Dean, College of Agricultural and Environmental Sciences, University of California, Davis

Eli Fromm
Vice Provost for Research and Graduate Studies, Drexel University

Jeff Froyd
Professor of Electrical Engineering, Rose-Hulman Institute of Technology

Thomas Fuja
Assistant Professor of Electrical Engineering, College of Engineering, University of Maryland

Robert A. Furgason
President, Accreditation Board for Engineering and Technology
President, Texas A&M University at Corpus Christi

Elsa M. Garmire
William Hogue Professor of Electrical Engineering, Director, Center for Laser Studies, University of Southern California

B. John Garrick
President, Chairman and CEO, PLG, Inc.

David B. Geselowitz
Professor of Bioengineering and Medicine, Pennsylvania State University

Mohammed S. Ghausi
Dean of Engineering, University of California, Davis

Reza Ghodssi
Graduate Electrical Engineering Student, Department of Electrical & Computer Engineering, University of Wisconsin–Madison

Joseph H. Gibbons
Associate Dean of Engineering, University of South Carolina

Norman A. Gjostein
Director, Materials Research Laboratory, Ford Motor Company Scientific Research Laboratory

Mario J. Gonzalez
Associate Vice Chancellor, South Texas/Border Area Development
Temple Foundation Endowed Professor, University of Texas System

James R. Goodman
Vice President for Academic Affairs, South Dakota School of Mines and Technology

Richard E. Goodman
Professor of Geological Engineering, Department of Civil Engineering, University of California, Berkeley

Jerrier A. Haddad
Retired Vice President, IBM Corporation
President-Elect, Consultant, Accreditation Board for Engineering and Technology

Carl W. Hall
Engineering Information Services

Richard F. Hartung
Executive Vice President, LOCKHEED IMS—a Subsidiary of Lockheed Corporation

E.R. (Vald) Heiberg III
President, J.A. Jones Construction Services Company

Mogens Hendriksen
Dean, School of Engineering and Mines, University of North Dakota

Peter G. Hoadley
Chair-Elect, ASCE Education Activities Committee, Professor of Civil and Environmental Engineering, Vanderbilt University

David A. Hodges
Professor of Electrical Engineering and Computer Sciences, Dean of Engineering, University of California, Berkeley

Joseph A.C. Humphrey
Professor, Department of Mechanical Engineering, University of California, Berkeley

Gary T. Hurford
President, Hunt Oil Company

Peggy Johnson
Assistant Professor of Civil Engineering, College of Engineering, University of Maryland

J.B. Jones
Randolph Professor Emeritus, Department of Mechanical Engineering, Virginia Polytechnic Institute and State University

David T. Kao
Dean of Engineering, College of Engineering, Iowa State University

W. David Kingery
Regents Professor, University of Arizona

H. Allan Knappenberger
Associate Dean, College of Engineering, Wayne State University

Edmond I. Ko
Professor, College of Engineering, Carnegie Mellon University

Caryn G. Korshin
Program Officer, Exxon Education Foundation

John L. Jackson, Jr.
Chairman and CEO, INDRESCO, Inc.

Jay Labov
Director, Committee on Undergraduate Science Education, National Research Council

Kenneth R. Laker
IEEE Vice President of Educational Activities, School of Engineering and Applied Science, University of Pennsylvania

James S. Langer
Director, Institute for Theoretical Physics, University of California, Santa Barbara

Kathryn Laskey
Associate Professor, Systems Engineering, School of Information Technology and Engineering, George Mason University

Allen Latham, Jr.
Founder, Haemonetics Corporation

George C. Lee
Dean, School of Engineering and Applied Sciences, State University of New York at Buffalo

Peter Y. Lee
Dean, School of Engineering, California Polytechnic State University

Thomas H. Lee
President, Center for Quality Management, Cambridge, Massachusetts

Joseph F. Lestingi
Dean, School of Engineering, The University of Dayton

Stanley R. Liberty
Dean of Engineering and Technology, University of Nebraska at Lincoln

Peter W. Likins
President, Lehigh University

Howard M. Liljestrand
Professor, Graduate Student Advisor, Raymond F. Dawson Centennial Teaching Fellowship in Engineering, Department of Civil Engineering, University of Texas at Austin

Arthur Linkins
Vice President for Academic Affairs, Clarkson University

Isabel Lloyd
Assistant Professor of Materials & Nuclear Engineering, College of Engineering, University of Maryland College Park

John D. Mackenzie
Nippon Sheet Glass Professor of Materials Science, Associate Dean, School of Engineering and Applied Science, University of California, Los Angeles

Harold Lee Martin
Dean of Engineering, School of Engineering, North Carolina Agricultural and Technical State University

Frank A. McClintock
Professor of Mechanical Engineering, Massachusetts Institute of Technology

John J. McCoy
Dean of Engineering, Catholic University of America

Mike McGrath
Program Director, Scientific Computing Directorate for Computer & Information, Science and Engineering, National Science Foundation

Kenneth G. McKay
former Executive Vice President, AT&T Bell Laboratories
Consultant

Eugene S. Meieran
Component Technology and Development Group, Intel Corporation

David G. Messerschmitt
Professor of Electrical Engineering and Computer Sciences, University of California, Berkeley

Ioannis Minis
Assistant Professor of Mechanical Engineering, College of Engineering, University of Maryland, College Park

Carl L. Monismith
The Robert Horonjeff Professor of Civil Engineering, Department of Civil Engineering, University of California, Berkeley

Walter P. Moore, Jr.
Chairman and President, Walter P. Moore & Associates, Inc.

Alfred Moye
Manager, Engineering Education, Hewlett-Packard Company

Tom Mulinazzi
Associate Dean, School of Engineering, University of Kansas

Godfrey Mungal
Associate Professor, Mechanical Engineering Department, Stanford University

E. Phillip Muntz
M.C. Gill Professor of Composite Materials and Chairman, Department of Aerospace Engineering, University of Southern California

Mark B. Myers
Vice President, Research, Xerox Corporation

Larisa Naples
Student, Engineering and Public Policy, Carnegie Mellon University

Venkatesh Narayanamurti
Dean of Engineering, University of California, Santa Barbara

Arnold W. Oliver
Executive Director, Texas Department of Transportation

Charles R. O'Melia
Professor of Environmental Engineering, Johns Hopkins University

Richard Opsahl
Grumman, Inc.

Simon Ostrach
Wilbert J. Austin Distinguished Professor of Engineering, Department of Mechanical and Aerospace Engineering, Case Western Reserve University

John Patterson
Dean of Engineering, School of Engineering, Manhattan College

David L. Pehrson
Deputy Associate Director, Engineering & Technology Transfer, Lawrence Livermore National Laboratory

John B. Peller
President and Center Director, Systems Development Center, Rockwell International

Alan W. Pense
Vice President and Provost, Lehigh University

Courtland D. Perkins
Retired President, National Academy of Engineering Consultant

Thomas K. Perkins
Distinguished Research Advisor, ARCO Exploration and Production Technology

Winfred Phillips
Chairman, Engineering Deans Council; Dean, College of Engineering, University of Florida

Karl S. Pister
BEEd Chair
Chancellor, University of California, Santa Cruz

Kerstin B. Pollack
Acting Director, Board on Engineering Education, National Research Council

Tom Price
Manager, Washington Office, Society of Automotive Engineers

C.E.G. Przirembel
Professor, Department of Mechanical Engineering, Clemson University

Martin R. Ramirez
Professor, Department of Civil Engineering, Johns Hopkins University

L.M. (Mike) Rice, Jr.
President and Member of the Board, Texas Instruments Foundation

Robert W. Ritchie
Director, University Affairs, Hewlett-Packard Company

T. Richard Robe
Dean of Engineering and Technology, Ohio University

John P. Robinson
Associate Dean of Engineering, University of Iowa

Decatur B. Rogers
Dean of Engineering, College of Engineering and Technology, Tennessee State University

Joseph E. Rowe
Associate Vice President for Research and Director, Research Institute, University of Dayton

Paul E. Rubbert
Unit Chief, Aerodynamics Research, Boeing Commercial Airplane Group

T.W. Fraser Russell
Director, Institute of Energy Conversion, University of Delaware

William Salmon
Executive Officer, National Academy of Engineering

Mario G. Salvadori
Honorary Chairman, Salvadori Educational Center, Weidlinger Associates, Consulting Engineers, P.C.

Lee W. Saperstein
Dean, School of Mines and Metallurgy, University of Missouri at Rolla

Frank D. Schowengerdt
Dean of Engineering, Colorado School of Mines

Frank J. Schuh
President, Drilling Technology, Inc.

Robert S. Sechler
Manager, Education Relations Department, Society of Automotive Engineers

Maurice M. Sevik
Head, Signatures Directorate, David Taylor Model Basin, Carderock Division, Naval Surface Warfare Center

Yatish T. Shah
Dean of Engineering, Drexel University

Fredrick H. Shair
Manager, Educational Affairs Office, Jet Propulsion Laboratory

Yazan N. Sharif
President, INDRESCO, Inc.

David P. Shattuck
Associate Dean of Engineering for Undergraduate Programs, University of Houston

Israel Sheinberg
President, Sheinberg Associates

R.D. Shelton
Chair, Engineering Department, Loyola College

Merrill I. Skolnik
Superintendent, Radar Division, Naval Research Laboratory

David Sloan
Assistant Vice President, Reid Investments

Theoren P. Smith, III
Director of Physical Sciences, Thomas J. Watson Research Center, IBM

James J. Solberg
Professor of Industrial Engineering, School of Industrial Engineering Director, Engineering Research Center, Purdue University

Mason H. Somerville
Dean of Engineering and Technology, Texas Tech University

John Steadman
President, National Council of Examiners for Engineering and Surveying

William D. Stevens
Retired Chairman, Foster Wheeler Corporation

Warren Stevenson
Assistant Dean, Schools of Engineering, Purdue University

H. Guyford Stever
Trustee and Advisor

Victor Szebehely
Professor and R.B. Curran Chair, Department of Aerospace Engineering and Engineering Mechanics, University of Texas at Austin

Bill M. Thompson
Retired Vice Chairman, Phillips Petroleum Company

James M. Tien
Acting Dean, School of Engineering, Rensselaer Polytechnic Institute

J. Howard Todd
Vice President, Engineering, Dupont Company

Alba Torrents
Assistant Professor of Civil Engineering, College of Engineering, University of Maryland College Park

Myron Tribus
Director, Exergy, Inc.

Marshall P. Tulin
Director, Ocean Engineering Laboratory, University of California, Santa Barbara

George L. Turin
Vice President, Teknekron Corporation

George Van Dusen
Assistant Dean of Engineering, College of Engineering, Michigan State University

Charles M. Vest
President, Massachusetts Institute of Technology

Oswald G. Villard, Jr.
Senior Scientific Advisor, SRI International

M. Lucius Walker, Jr.
Dean, School of Engineering, Howard University

William F. Walker
Dean of Engineering, Auburn University

Don Weinert
Executive Director, National Society of Professional Engineers

Norman Wereley
Assistant Professor of Aerospace Engineering, College of Engineering, University of Maryland, College Park

Paul Wesling
Advisory Design Engineer, Tandem Computers, Inc.

Chelsea C. White, III
Senior Associate Dean, College of Engineering, University of Michigan

E. Bernard White
Assistant Dean for Undergraduate Studies, School of Information Technology and Engineering, George Mason University

Edward J. White
Associate Dean, School of Engineering, Vanderbilt University

Robert M. White
President, National Academy of Engineering

Robert V. Whitman
Professor Emeritus of Civil Engineering, Massachusetts Institute of Technology

F. Karl Willenbrock
Consultant

Frank L. Williams
Dean of Engineering, University of Alaska

Robert F. Witte
Program Officer, Exxon Education Foundation

Jack Keil Wolf
Stephen O. Rice Professor of Electrical and Computer Engineering, Center for Magnetic Recording Research, University of California, La Jolla

Archie L. Wood
Executive Director, Commission on Engineering and Technical Systems, National Research Council

David N. Wormley
Dean of Engineering, Pennsylvania State University

Robert E. Wulf
Vice President, Engineering and Technology, Northrop Corporation

William A. Wulf
AT&T Professor of Engineering and Applied Science, Department of Computer Science, University of Virginia

Lan Xue
Department of Engineering Management, School of Engineering and Applied Science, George Washington University

Yi-Yuan Yu
Professor Emeritus, Mechanical Engineering, New Jersey Institute of Technology

Dorothy S. Zinberg
Lecturer, Public Policy Center for Science and International Affairs, Kennedy School of Government, Harvard University

Appendix C

Contributors to the Study

August 22-23, 1991. Board Colloquium on "Culture Change in Engineering Education"

November 19, 1991. Workshop on "Internationalization of Engineering Practice: Changing Roles for Education"

January 30-31, 1992. Board Colloquium on "Outreach in Engineering Education"

May 18-19, 1992. Board Colloquium on "Innovative Models for Engineering Education and Related Methods for Achieving Culture Change"

July 21-22, 1992. Workshop on "Emerging Technologies and Their Impact on the Delivery of Engineering Education"

September 10-11, 1992. Board Colloquium on "The Urgency of Outreach to Underrepresented Minorities and Women in Engineering Education"

February 4-5, 1993. Board Colloquium on "Diminishing Resources and Changing Missions for Engineering Education: Options for Action"

Individuals in bold typeface made presentations or participated in panel discussions; the rest contributed to the general discussion.

FEDERAL GOVERNMENT

Department of Defense

Office of the Secretary of Defense

- **Victor Reis**, Director of Defense Research and Engineering
- Leo Young
- Jasper Lupo
- Fred E. Saalfeld, Office of Naval Research

Office of the Assistant Secretary of Defense for Force Management and Personnel

- Carl Dahlman, Deputy Assistant Secretary of Defense for Requirements and Resources
- Janet Johnston, Research Specialist
- Judy Fernandez, Economist

Department of Education

- Clifford Adelman, Director, Higher Education, Office of Research

Department of Energy

- Richard E. Stephens, Associate Director, Office of University and Science Education, Office of Energy Research

Department of Labor

- **Ronald E. Kutscher**, Associate Commissioner, Office of Employment Projections, Bureau of Labor Statistics

House Subcommittee on Science, Research and Technology, U.S. Congress

- Robert A. Ellson, ASME Congressional Fellow

National Aeronautics and Space Administration

- Elaine Schwartz, Chief, University Programs Branch
- Richard Devon, Associate Director, National Space Grant College and Fellowship Program

National Science Foundation

National Science Board

- **Jaime Oaxaca**, Member; Vice Chairman, Coronado Communications Corporation

Directorate for Education and Human Resources

- **Luther Williams**, Director

Office of Studies, Evaluation, and Dissemination

- Kenneth J. Travers, Office Head

Division of Undergraduate Science, Engineering & Mathematics Education

- **Robert F. Watson**, Division Director
    - Jacob M. Abel, Program Director
    - James G. Harris, Program Director
    - Harry Hedges, Program Director
    - Doris K. Lidtke, Program Director
    - Norman L. Fortenberry, Program Director
    - Chalmers Sechrist, Program Director
    - George D. Peterson, Section Head and Program Director

Directorate for Engineering

- **Joseph Bordogna**, Assistant Director
- Ray Bowen, Deputy Assistant Director

Division of Engineering Infrastructure Development

- Irene Peden, Director
- **Susan Kemnitzer**, Deputy Director
- Win Aung, Senior Staff Associate
- Frank (Dale) Draper, Senior Staff Associate
- Lucy Morse, Program Manager, Human Resources and Education

Division of Engineering Centers

- **Marshall Lih**, Director
- Lynn Preston, Deputy Director

Division of Design and Manufacturing Systems

- **F. Hank Grant**, Acting Program Director, Design & Computer-Integrated Engineering Program, and Program Director, Operations Research & Production Systems; Head, Task Force on Quality in Engineering Education

Directorate for Science, Technological, and International Affairs

- **F. Karl Willenbrock**, Assistant Director

Directorate for Social, Behavioral and Economic Science

- **Kenneth M. Brown**, Division Director, Division of Science Resources Studies

Office of Science and Technology Policy

- Charles Dickens, Senior Staff Associate, FCCET
- Arthur Sheekey, Fellow
- **Eugene Wong**, Associate Director for Industrial Technology

ENGINEERING EDUCATION CONSTITUENCIES

Accreditation Board for Engineering and Technology (ABET)

- Leslie Benmark, President
- **Edward W. Ernst**, Past President (Allied Signal Professor of Engineering, University of South Carolina)
- A.T. Kersich, President-Elect; HKM Associates
- **John W. Prados**, Past President
- David Reyes-Guerra, Executive Director

American Association of Engineering Societies (AAES)

- Lawrence P. Grayson, Past Chairman

American Association of Higher Education

- Clara M. Lovett, Director, Forum on Faculty Reward System

American Institute of Chemical Engineers

- Betsy Houston

American Society for Engineering Education (ASEE)

- Leighton E. Sissom, President
- Frank Huband, Executive Director
- Ann Leigh Speicher, Manager, Federal Liaison

American Society of Civil Engineering (ASCE)

- Louis L. Guy, Jr., Secretary, Committee on Curriculum and Accreditation
- Luther Graef, President, Graef, Anholt, and Schloemer; member ASCE Board of Directors; Director, ASCE Accreditation Board for Engineering and Technology)
- Delon Hampton, Director, ASCE District 5 (President, Delon Hampton & Associates)

American Society of Mechanical Engineers (ASME)

- Winfred Phillips, Senior Vice President and Chairman of ASME Council on Education; Dean, College of Engineering, University of Florida
- Chor Tan, ASME Managing Director of Education; formerly Dean of School of Engineering at Cooper Union

Engineering Deans' Council (EDC)

- Eleanor Baum, Chair
- Frank Huband, Executive Director

Institute of Electrical and Electronic Engineers (IEEE)

- Martha Sloan, President
- Eric E. Sumner, President
- Eric Herz, General Manager
- Rudolph A. Stampfl, Staff Director, Educational Activities
- M. E. Van Valkenburg, Dean Emeritus of Electrical and Computer Engineering, University of Illinois; representing IEEE Educational Activities Board

National Action Council for Minorities in Engineering, Inc.

- **George Campbell, Jr.**, President National Society of Professional Engineers (NSPE)
- Frank A. Kulacki, President-Elect
- Neil A. Norman, President
- Donald G. Weinert, Executive Director

Society of Manufacturing Engineers

- Frank H. McCarty, President-elect
- Frank Riley, President, Manufacturing Engineering Education Foundation
- Keith Bankwitz, Manager, Manufacturing Engineering Education Foundation
- Fred Michel, Member, SME Government Relations Committee; Chairman, CASA/SME Board of Advisors

ACADEME

Big 10+ Schools of Engineering

University of California, Berkeley

- **David A. Hodges**, Dean of Engineering
- Alice Agogino, Coalition Associate Director, Curriculum, Synthesis (A National Engineering Education Coalition)

Cornell University

- **William B. Streett**, Dean of Engineering

Georgia Institute of Technology

- **John A. White**, Dean of Engineering
- **Jack R. Lohmann**, Industrial Engineering
- **Norman Johnson**, Executive Assistant to the President

University of Michigan

- **Peter M. Banks**, Chairman of Deans Group

Michigan State University

- **Nicholas Aterio**, Associate Dean for Research and Graduate Studies of the College of Engineering

University of Minnesota

- **Gordon Beavers**, Dean of Engineering

Northwestern University

- **Stephen H. Carr**, Associate Dean, McCormick School of Engineering and Applied Science

Ohio State University

- **Robert F. Redmond**, Associate Dean of Engineering Emeritus
- Stacy Weislogel, Associate Dean, College of Engineering

The Pennsylvania State University

- **Carl Wohlgemuth**, Associate Dean of Engineering Emeritus

Purdue University

- **Henry T. Yang**, Dean of Engineering
- **James Solberg**, Director, Engineering Research Center for Intelligent Manufacturing Systems

Stanford University

- **James F. Gibbons**, Dean of Engineering

The University of Texas at Austin

- **Herbert H. Woodson**, Dean of Engineering

University of Wisconsin–Madison

- **John G. Bollinger**, Dean of Engineering
- William W. Wuerger, College of Engineering

Historically Black Colleges and Universities

Florida A&M/Florida State University

- Ching-Jen Chen, Dean

Howard University

- Mobolaji E. Aluko, Associate Director, Pipeline, ECSEL; Chemical Engineering Department, School of Engineering
- Bruce Schimming, Administrative Director, School of Engineering
- **M. Lucius Walker, Jr.**, Dean of Engineering, School of Engineering; Principal Investigator, ECSEL

North Carolina Agricultural and Technical University

- **Harold Martin**, Dean of Engineering
- John C. Kelly, Jr., Associate Dean

Other Colleges and Universities

Arizona State University

- David C. Chang, Dean, College of Engineering and Applied Sciences

Auburn University

- William F. Walker, Dean of Engineering

University of California, San Diego

- **M. Lea Rudee**, Dean of Engineering, Division of Engineering

Carnegie Mellon University

- Stephen W. Director, Dean, Carnegie Institute of Technology

Catholic University of America

- John J. McCoy, Dean, School of Engineering
- Timothy Kao

University of Cincinnati

- Constantine N. Papadakis, Dean, College of Engineering

University of Colorado at Boulder

- John Dunn, Associate Dean, Undergraduate Programs

University of Colorado at Denver

- Peter E. Jenkins

Colorado School of Mines

- Michael B. McGrath

Colorado State University

- **Frank A. Kulacki**, Dean of Engineering, College of Engineering

Columbia University

- David H. Auston, Dean of Engineering
- **Mischa Schwartz**, Professor of Electrical Engineering, Center for Telecommunications Research
- John R. Kender, Computer Science

Community College of Baltimore

- **James D. Tschechtelin**, President

University of Delaware

- Dan Boulet, Assistant Dean of Engineering, College of Engineering

Drexel University

- **Dennis Brown**, Provost
- **Eli Fromm**, Vice Provost of Research and Graduate Studies
- Thomas Canavan, Dean of Arts and Sciences
- **Robert Quinn**, Professor of Electrical and Computer Engineering
- **Y. T. Shah**, Dean of Engineering
- Project Team Leaders:
  Valarie Arms, Robin Carr, Michael Gealt, Alan Lawley, Donald Larson, Tom Moore, Allan Smith, Mrat Tanyel, Charles Weinberger
- Participating students:
  Keith Alyea, Adam Bair, Scott Bussell, Jeff Catteau, Bill Christine, Tim Coffman, George Dixey, Shital Godhania, Mike Goslin, Don Haring, Kevin Heinsey, Matt Hudson, Bob Kosto, Rachel Leonard, Rebekah Rupp, Todd Springer, Timothy Warren, Brian Woolf

Duke University

- Earl Dowell, Dean of Engineering

Florida Institute of Technology

- Robert L. Sullivan, Dean, College of Engineering

The George Washington University

- Gideon Frieder, Dean, School of Engineering

University of Houston

- David P. Schattuck

Iowa State University

- **David T. Kao**, Dean of Engineering, College of Engineering

The Johns Hopkins University

- **Ross Corotis**, Associate Dean for Academic Affairs, G. W. C. Whiting School of Engineering

University of Kansas

- Carl E. Locke, Jr., Dean of Engineering

University of Maryland

- George E. Dieter, Jr., Dean of Engineering
- **Thomas M. Regan**, Associate Director, Teaching/Learning Innovation Program, ECSEL (Engineering Coalition of Schools for Excellence in Education and Leadership); Professor and Associate Chair, Department of Chemical Engineering

University of Massachusetts

- Keith R. Carver, Acting Dean

Massachusetts Institute of Technology

- **Thomas L. Magnanti**, Co-director, Leaders for Manufacturing Program; George Eastman Professor of Management Science
- **David Wormley**, Associate Dean; Professor of Mechanical Engineering

MESA Statewide Program

- **Wilfred (Fred) O. Easter, Jr.**, Executive Director

University of Miami

- Martin Becker, Dean, College of Engineering

Mississippi State University

- **Robert A. Altenkirch**, Dean of Engineering, College of Engineering

University of Missouri–Columbia

- **Paul Braisted**, Assistant Dean of Instructions, College of Engineering

Montana State University

- David F. Gibson, Dean, College of Engineering

National Consortium for Graduate Degrees for Minorities in Engineering and Science, Inc. (GEM)

- **Howard G. Adams**, Executive Director, The National Center for Graduate Education for Minorities

National Technological University

- **Herbert Rabin**, member, Board of Trustees; Director, Engineering Research Center, University of Maryland
- Donald B. Coffin, Senior Project Manager

New Jersey Institute of Technology

- George Pincus, Dean of Engineering, Newark College of Engineering

University of New Mexico

- James K. Thompson, Dean, College of Engineering

North Carolina State University

- Wilbur L. Meier, Jr., Dean of Engineering

Northeastern University

- Paul H. King, Dean, College of Engineering

University of Oklahoma

- Billy L. Crynes, Dean, College of Engineering

Polytechnic University

- Nancy M. Tooney, Assistant Provost for Academic Affairs

University of Puerto Rico

- **Jose F. Lluch**, Dean of Engineering, College of Engineering

University of Rhode Island

- Thomas J. Kim, Interim Dean, College of Engineering

Rose-Hulman Institute of Technology

- **Samuel F. Hulbert**, President
- **James R. Eifert**, Dean of Engineering

San Jose State University

- **Jay D. Pinson**, Dean of Engineering, School of Engineering

Santa Clara University

- **Terry E. Shoup**, Dean of Engineering, School of Engineering

South Dakota State University

- Duane Sander, Dean of Engineering

University of Southern California

- Leonard M. Silverman, Dean, School of Engineering

State University of New York at Buffalo

- George C. Lee, Dean, School of Engineering & Applied Sciences

The University of Texas at San Antonio

- **Manuel Berriozabal**, Professor of Mathematics; Coordinator of TexPREP

Texas A&M University

- **Kenneth Lee Peddicord**, Dean of Engineering, College of Engineering

Texas Tech University

- Mason Somerville, Dean, College of Engineering

University of Toledo

- Edward Lumsdaine, Dean, College of Engineering

Tulane University

- William C. Van Buskirk, Dean of Engineering

University of Virginia

- Edgar A. Starke, Jr., Dean of Engineering and Applied Science

Washington University

- **Christopher I. Byrnes**, Dean of Engineering, School of Engineering and Applied Science

OTHER INSTITUTIONS

Alfred P. Sloan Foundation

- **Hirsch Cohen**, Program Officer

Allied Signal Inc.

- **Mary L. Good**, Senior Vice President-Technology

American Association for the Advancement of Science

- **Shirley M. Malcom**, Head, Directorate for Education and Human Resources
- **Marsha Lakes Matyas**, Project Director, Directorate for Education and Human Resources

American Electronics Association

- **Eve C. Majure**, Manager, University Programs, Education and Science Policy,

American Cyanamid Company

- **Edward W. Cantrall**, Vice President of Operations, Medical Research Division

Commission on Professionals in Science and Technology

- **Betty M. Vetter**, Executive Director

Council on Competitiveness

- **Erich Bloch**, Distinguished Fellow

Digital Equipment Corporation

- Janet Potter

GE Corporate Research and Development Center

- **Walter L. Robb**, Senior Vice President (retired), Corporate R&D

Howmet Corporation

- **Neil Paton**, Vice President-Technology

Industrial Research Institute

- Charles F. Larson, Executive Director
- Robert E. Burkhart

Moog Corporation

- **Richard Aubrecht**, Chief Executive Officer

National Center for Manufacturing Sciences

- **Edward Miller**, President
- Rendard Gueringer, Manager, Management Practices

SRI International

- James O. Gollub, Director, The Center for Science and Engineering Education Development (SEED)

Xerox Corporation

- **Esther M. Conwell**, Senior Research Fellow

ACADEMY COMPLEX

National Academy of Engineering

- William C. Salmon, Executive Officer

National Research Council

Committee on Post-secondary Education and Training for the Workplace

- **Barry Sugarman**, Staff Officer

Coordinating Council for Education

- **Kenneth M. Hoffman**; NRC Associate Executive Officer for Education

Mathematical Sciences Education Board

- **Ray C. Shiflett**, Executive Director
- Susan L. Forman, Director of College and University Programs

Commission on Physical Sciences, Mathematics and Applications

Board on Mathematical Sciences

- John J. Lavery, Staff Director

Commission on Engineering and Technical Systems

Liaison Members to BEEd

- Dennis Chamot, Associate Director, Department of Professional Employees, AFL-CIO
- James J. Solberg, Professor of Industrial Engineering, Purdue University
- Paul Torgersen, President, Corporate Research Center, College of Engineering, Virginia Polytechnic Institute and State University

Executive Office

- Archie L. Wood, Executive Officer
- Marlene Beaudin, Assoc. Exec. Officer

Office of Japan Affairs

- Martha C. Harris, Director

Office of Scientific and Engineering Personnel

Liaison Member to BEEd

- John P. Crecine (represented by John A. White, Dean of Engineering Georgia Institute of Technology)

Executive Office

- Claudia Dissel, Associate Executive

Director

- Alan E. Fechter, Executive Director

- Pamela E. Flattau, Director, Surveys and Studies, OSEP
- James Voytuk, Project Director, Doctoral and Postdoctoral Study in the Mathematical Sciences (on loan to BEEd)

Committee on Women in Science and Engineering

- **Linda Dix**, Study Director

Appendix D

Toward A Progressive New Engineering Curriculum

The following outline, prepared by the Board on Engineering Education (BEEd), represents a description of the purpose and principles of a new curriculum, as well as the means and mechanisms required to advance it on a nationwide basis. It is presented as an example of the type of curriculum that is suitable for contemporary engineering education. Each institution is urged to develop its own curriculum tailored to its strengths, its student population, and its own vision of future needs. It is important, however, for each institution to maintain an awareness of curriculum development activities that are ongoing nationwide and to adopt those developments that are applicable to its circumstances.

PURPOSE

The purpose of undergraduate engineering education is threefold. First, it provides a course of study that prepares students to enter the practice of engineering in a selected professional field. Second, it must be broad enough to provide a strong basis not only for a career in engineering but also for careers in other professions, in business, or in public service. (This broad purpose makes engineering education a prime vehicle for encouraging wider participation by women and minority students in engineering, and it broadens the opportunities for employment of graduates during periods of limited hiring of engineers.) Third, it must make students aware of the relationships between engineering and industrialized society, encouraging and preparing them to assume a stronger and more visible leadership position as engineers in society and as productive citizens.

PRINCIPLES

To accomplish this purpose, the new undergraduate engineering curricula and culture should:

1. Provide broad, solid knowledge of key fundamental concepts in science and engineering. These concepts should not be taught only in the abstract but also with constant reference to engineering practice.
2. Provide in-depth engineering study in at least one field. Part of this study should address business and management aspects in that field of engineering and encompass a focus on global practice—some of which may be captured in a capstone design project.
3. Incorporate the study of engineering practice within the curriculum. This includes opportunities for "apprenticeship," possibly through co-op or summer internship programs. It also implies that some faculty will make studies of practice a part of their scholarship.
4. Provide an ability to understand the major societal issues and to lead in addressing them through technology.
5. Provide greater flexibility to pursue other careers outside engineering.
6. Impart an ability to learn independently.
7. Establish a new culture and a new image for engineering education and practice that is humane and that will attract and retain students, with a particular focus on women and underrepresented minorities.

MEANS

Suggested means to achieve these principles of the new undergraduate engineering curriculum are

1. Integrate new material and different perspectives into the undergraduate engineering curriculum using approaches such as:

- offering a first-year course on the transformation of society by engineering, giving concrete examples;
- introducing engineering illustrations into the mathematics and science courses taken by engineering students;
- introducing case studies into the engineering science courses to illustrate how the principles of the subject, including their math and science roots, have developed;
- encouraging a better integration of liberal arts into the curriculum;
- increasing the emphasis, especially in upper-division courses, on engineering practice through the study of contemporary innovations and problems; and

• introducing a senior thesis/project in which students research an idea and defend it in a detailed written text or design and build a prototype for a novel engineered system.

2. Integrate into the curriculum a number of important concepts, such as:

• enjoyment and fun in the learning process;
• design experience;
• team research/design experience, with oral reporting by teams;
• academic study of engineering practice;
• globalization of technology, understanding other cultures, and appreciation of the liberal arts;
• exposure to the concepts of business, economics, marketing, and manufacturing, and risk;
• sustainable development of the environment; and
• engineering management, including effective interaction with shop-floor and technical support personnel.

3. Develop activities that help broaden the student's outlook and experience and that are synergistic with the curriculum (e.g., activities that involve technology and politics, technology and religion, or technology and art).

4. Remove some material and some courses from the current curriculum. If the curriculum is to remain manageable and able to be completed within the current timeframe of four years, it is important that the curriculum emphasize subjects of a fundamental nature and those that are more difficult for students to learn on their own, such as engineering design. Remove redundancies, for example, the repetitious teaching of the same principles of chemistry, physics, and thermodynamics in different courses. Incorporate some math and science "base" courses into engineering courses. Emphasize in-depth one area of engineering practice in a discipline and provide a broad overview of other areas—for example, in manufacturing engineering emphasize robotics and provide an overview of process simulation, materials handling, etc. Ensure that students in a given discipline have at least some familiarity with other engineering disciplines; multidisciplinary capstone design projects can help.

5. Go to a five-year curriculum, with the first four years providing a broad bachelor of science degree and the fifth year leading to an in-depth professional specialization degree. (Obviously, such a curriculum adds considerably to the cost of an engineering degree and the time required to complete it. For that reason, past experiments with a five-year program have not been highly successful.)